# Springer Monographs in Mathematics

For other titles published in this series, go to
www.springer.com/series/3733

Susan Barwick · Gary Ebert

# Unitals in Projective Planes

 Springer

Susan Barwick
Pure Mathematics
The University of Adelaide
Adelaide, Australia
susan.barwick@adelaide.edu.au

Gary Ebert
Department of Mathematics
The University of Delaware
Newark, DE, USA
ebert@math.udel.edu

ISSN: 1439-7382
ISBN: 978-1-4419-2619-7          e-ISBN: 978-0-387-76366-8
DOI: 10.1007/978-0-387-76366-8

Mathematics Subject Classification (2000): 51Axx:51A35

Printed on acid-free paper

springer.com

This book is dedicated to
Rey Casse
and in memory of Richard Bruck

# Preface

This book is a monograph on unitals embedded in finite projective planes. Unitals are an interesting structure found in square order projective planes, and numerous research articles constructing and discussing these structures have appeared in print. More importantly, there still are many open problems, and this remains a fruitful area for Ph.D. dissertations.

Unitals play an important role in finite geometry as well as in related areas of mathematics. For example, unitals play a parallel role to Baer subplanes when considering extreme values for the size of a blocking set in a square order projective plane (see Section 2.3). Moreover, unitals meet the upper bound for the number of absolute points of any polarity in a square order projective plane (see Section 1.5). From an applications point of view, the linear codes arising from unitals have excellent technical properties (see Section 6.4). The automorphism group of the classical unital $\mathcal{H} = \mathcal{H}(2, q^2)$ is 2-transitive on the points of $\mathcal{H}$, and so unitals are of interest in group theory. In the field of algebraic geometry over finite fields, $\mathcal{H}$ is a maximal curve that contains the largest number of $\mathbb{F}_{q^2}$-rational points with respect to its genus, as established by the Hasse-Weil bound.

The book is a thorough survey of the research literature on embedded unitals, the first time such material has been collected and presented in one place. The intended audience consists of graduate students and mathematicians who want to learn this subject area without reading all the original articles. Moreover, the book should serve as a useful reference for researchers already working in the area. The primary proof techniques used involve linear algebraic arguments, finite field arithmetic, some elementary number theory, and combinatorial enumeration. Some computer results not previously found in the literature are mentioned in the text.

We assume that the reader has some familiarity with projective planes, and cover the necessary preliminary material in Chapter 1. In Chapter 2 we introduce Hermitian curves in $\mathrm{PG}(2, q^2)$ and study their geometric and combinatorial properties. We then define unitals as a generalization of Hermitian curves, and discuss unitals embedded in projective planes as well as unitals treated independently as designs. We note that while most unitals are not embedded in projective planes, the most interesting ones from a structural point of view are the embedded ones. Blocking sets and their relation to unitals are discussed, providing additional motivation for the study of embedded unitals.

Chapter 3 introduces translation planes and covers in detail some projective geometry techniques that are very useful in the study of unitals. Namely, we define the Bruck-Bose representation in $PG(4, q)$ for a translation plane of dimension at most two over its kernel. This representation is useful in a much broader context than the study of unitals, and the detailed presentation given here should prove to be invaluable for graduate students working in any area of finite geometry. This naturally leads to a discussion of the construction of spreads in $PG(3, q)$. Finally, the substructure of square order planes, and of Baer subplanes in particular, then leads to the notion of derivation as a means for constructing new projective planes from old ones.

Chapter 4 begins with Buekenhout's two constructions for embedded unitals. The first produces unitals in any translation plane of dimension at most two over its kernel, and the second produces unitals in certain derivable two-dimensional translation planes. It turns out that the second procedure produces only the classical unital (Hermitian curve) in the Desarguesian plane $PG(2, q^2)$, while the first procedure produces not only the classical unital but also numerous nonclassical unitals in $PG(2, q^2)$, as long as $q > 2$. The chapter concludes with a careful description of the known unitals embedded in $PG(2, q^2)$, all of which may be obtained from Buekenhout's first construction. Coordinates are given for these unitals, their structure is investigated, their groups are determined, and the number of nonequivalent unitals so constructed is determined.

Chapter 5 surveys unitals embedded in non-Desarguesian planes, beginning with a full investigation of the known unitals in the Hall plane $\text{Hall}(q^2)$. In particular, we discuss how these unitals are inherited from unitals in $PG(2, q^2)$ via derivation. Computer results are given for all possible unitals obtained from either of Buekenhout's two constructions when applied to Hall planes of small order. In particular, we note that there is always one unital which naturally arises from the first Buekenhout construction which is not obtained by any of the above derivation results. The chapter concludes with a survey of the known unitals in semifield planes, nearfield planes, Figueroa planes, and Hughes planes. Again some computer results are given for small orders.

In Chapter 6 we investigate various combinatorial properties of unitals and some associated configurations. We survey results concerning the intersection of a unital with a Baer subplane, as well as the intersection of two unitals. We look at spreads and packings of unitals, discuss the construction of inversive planes from certain unitals, investigate arcs contained in the classical unital, and look at the construction of codes from unitals.

In Chapter 7 a comprehensive survey is given of the known geometric characterizations of ovoidal-Buekenhout-Metz and classical unitals. Results concerning how many chords of a unital $U$ in $PG(2, q^2)$ need to be Baer sublines in order for $U$ to be classical or ovoidal-Buekenhout-Metz are surveyed and proved in detail. Other configurational characterizations are given, including Thas' proof of the longstanding conjecture that a unital $U$ in $PG(2, q^2)$ with collinear feet from every point in $PG(2, q^2) \setminus U$ must be classical. Finally, we briefly look at characterizations of unitals using the Bruck-Bose representation of $PG(2, q^2)$ in $PG(4, q)$ and the quadratic extension $PG(4, q^2)$, as well as characterizations using the Bose representation of $PG(2, q^2)$ in $PG(5, q)$.

The book concludes with Chapter 8, where a number of open problems are presented. Appendix A provides a standard naming system for the unitals arising from either of Buekenhout's two constructions, and includes a discussion of the conflicting names presently found in the literature. Appendix B catalogues and summarizes group theoretic characterizations of classical and ovoidal-Buekenhout-Metz unitals, whose group theoretic proofs are beyond the scope of this book.

The bibliography is a comprehensive list of research articles on unitals embedded in projective planes, and should be a valuable resource to anyone working in the area. A notation index is provided, as well as a conventional index of all terms used in the book.

The authors would like to thank the referees for valuable comments that improved the clarity and quality of the final version of this monograph. In addition thanks go to the editorial staff at Springer-Verlag for their assistance. Thanks to Rey Casse and Richard Bruck (in memoriam) for introducing us to the delights of finite geometry. Finally, we wish to thank our families for their patience and enduring support, without which this project would never have come to fruition.

*Susan Barwick*
*Gary Ebert*

March, 2008

# Contents

Preface ................................................................ VII

1  Preliminaries ...................................................... 1
   1.1  Affine and Projective Geometries ........................... 1
   1.2  Finite Fields .............................................. 8
   1.3  Quadrics in Low Dimensions ................................. 9
   1.4  Ovals and Ovoids .......................................... 12
   1.5  Some Linear Algebra ....................................... 13

2  Hermitian Curves and Unitals ..................................... 21
   2.1  Nondegenerate Hermitian Curves ............................ 21
   2.2  Degenerate Hermitian Curves and Baer Sublines ............. 24
   2.3  Unitals ................................................... 27

3  Translation Planes ............................................... 33
   3.1  Translation Planes ........................................ 33
   3.2  Derivation ................................................ 34
   3.3  Spreads ................................................... 36
   3.4  The Bruck-Bose Representation ............................. 41
        3.4.1  The Bruck-Bose Construction ........................ 41
        3.4.2  Baer Subplanes and Baer Sublines in Bruck-Bose ..... 43
        3.4.3  Derivation in Bruck-Bose ........................... 52
        3.4.4  Coordinates in Bruck-Bose .......................... 53

4  Unitals Embedded in Desarguesian Planes .......................... 59
   4.1  Buekenhout Constructions .................................. 59
   4.2  Unitals Embedded in $PG(2, q^2)$ .......................... 66
        4.2.1  The Odd Characteristic Case ........................ 66
        4.2.2  The Even Characteristic Case ....................... 79

**5   Unitals Embedded in Non-Desarguesian Planes** ............... 89
   5.1   Unitals in Hall Planes ..................................... 89
   5.2   Unitals in Semifield Planes ................................ 97
   5.3   Unitals in Nearfield Planes ................................ 101
   5.4   Unitals Embedded in Nontranslation Planes ................ 103
      5.4.1   Figueroa Plane .................................. 103
      5.4.2   Hughes Plane .................................... 105

**6   Combinatorial Questions and Associated Configurations** ........ 109
   6.1   Intersection Problems ..................................... 109
   6.2   Spreads and Packings ..................................... 117
   6.3   Related Combinatorial Structures .......................... 121
      6.3.1   Inversive Planes ................................. 121
      6.3.2   Arcs ........................................... 123
   6.4   Unitals and Codes ........................................ 127

**7   Characterization Results** ...................................... 133
   7.1   Characterizations of Unitals via Baer Sublines .............. 133
   7.2   Proofs of Results from Section 7.1 ......................... 136
   7.3   Other Configurational Characterizations ..................... 148
      7.3.1   Tallini Scafati Characterizations .................... 148
      7.3.2   Characterizations Using Feet ...................... 153
      7.3.3   Characterizations Using O'Nan Configurations ....... 161
   7.4   Characterizations Using the Quadratic Extension $PG(4, q^2)$ ... 164
   7.5   The Bose Representation of $PG(2, q^2)$ in $PG(5, q)$ ............ 164
   7.6   Group Theoretic Characterizations ......................... 166

**8   Open Problems** .............................................. 167

**A   Nomenclature of Unitals** ..................................... 171

**B   Group Theoretic Characterizations of Unitals** ................... 173

**References** ..................................................... 177

**Notation Index** ................................................. 187

**Index** ......................................................... 189

# 1

# Preliminaries

## 1.1 Affine and Projective Geometries

We begin by briefly recalling some fundamental definitions and ideas from affine and projective geometry. We assume that the reader has some familiarity with projective geometry, at least with real projective planes. Good references include Baer [18], Beutelspacher and Rosenbaum [46], Casse [85], Coxeter [90], Dembowski [96], Hall [128], and Hughes and Piper [140].

**Definition 1.1.** *An* **affine plane** $\mathcal{A}$ *is a set of objects, called points, together with certain subsets of points, called lines, such that*

1. *any two distinct points are contained in a unique common line,*
2. *given a point $P$ and a line $\ell$ with $P \notin \ell$, there exists a unique line $m$ with $P \in m$ and $\ell \cap m = \emptyset$,*
3. *there exist three points which are not contained in a common line.*

We often use geometric language such as "the point $P$ lies on the line $\ell$". If a subset of points lies on a common line, we say the points are **collinear**. Similarly, if a subset of lines all contain a common point, we say the lines are **concurrent**. Furthermore, if lines $\ell$ and $m$ have no common point, we say that $\ell$ and $m$ are **parallel**. Thus the second axiom above is typically called the **parallel axiom**. Note that the set of all lines in an affine plane can be partitioned into **parallel classes**. However, one must keep in mind that we are not assuming any notion of angle, distance, or any other metric property in our definition of affine plane.

*Example 1.2.* The Euclidean plane $\mathbb{R}^2$, stripped of its metric properties, is an affine plane.

*Example 1.3.* Let $V$ be a two-dimensional vector space over some field (or skew field) $F$. We consider $V$ as an additive group. Hence subspaces are (normal) subgroups, and cosets of a subspace are translates of that subspace.

S. Barwick, G. Ebert, *Unitals in Projective Planes*,
DOI: 10.1007/978-0-387-76366-8_1, © Springer Science+Business Media, LLC 2008

Take as *points* the vectors of $V$, including the zero vector, and take as *lines* the cosets of all one-dimensional vector subspaces of $V$. Then it is easy to see that this is an affine plane, which we will denote by $AG(2, F)$. If $F$ is a finite field, then $AG(2, F)$ is a finite affine plane.

When proving theorems about lines in affine planes, one often has to consider different cases, depending upon whether two lines meet or are parallel. This difficulty is resolved by completing the affine plane to a *projective plane*.

**Definition 1.4.** *A* **projective plane** $\mathcal{P}$ *is a set of objects, called points, together with certain subsets of points, called lines, such that*

  1. *any two distinct points are contained in a unique common line,*
  2. *any two distinct lines meet in a unique point,*
  3. *there exist four points, no three of which are contained in a common line.*

We adopt the same geometric language for projective planes as we did for affine planes. From the above three axioms for a projective plane, one can show that there exist four lines, no three of which are concurrent. Hence the axioms for a projective plane are *self-dual*, in the sense that one may interchange points and lines and thereby obtain another projective plane, called the **dual** of the original projective plane. This leads to the following result.

**Theorem 1.5 (Principle of Duality).** *If a theorem is valid for all projective planes, then the dual theorem obtained by interchanging the notions of point and line is also valid for all projective planes.*

However, one must be warned that a projective plane and its dual need not be *isomorphic*.

**Definition 1.6.** *Let* $\mathcal{P}_1$ *and* $\mathcal{P}_2$ *be projective planes. An* **isomorphism** *from* $\mathcal{P}_1$ *to* $\mathcal{P}_2$ *is a bijection* $\phi$ *from the points and lines of* $\mathcal{P}_1$ *to the points and lines of* $\mathcal{P}_2$ *that preserves containment. That is, a point $P$ lies on a line $\ell$ of* $\mathcal{P}_1$ *if and only if the point $P^\phi$ lies on the line $\ell^\phi$ of* $\mathcal{P}_2$. *If such a map exists, then* $\mathcal{P}_1$ *and* $\mathcal{P}_2$ *are called* **isomorphic***, denoted by* $\mathcal{P}_1 \cong \mathcal{P}_2$. *If* $\mathcal{P}_1 = \mathcal{P}_2$, *then an isomorphism from* $\mathcal{P}_1$ *to itself is called an* **automorphism** *(or* **collineation***) of* $\mathcal{P}_1$.

One should note that if $\phi$ is an isomorphism from $\mathcal{P}_1$ to $\mathcal{P}_2$ and if $\mathcal{L}_1$ denotes the lines of $\mathcal{P}_1$, then the lines of $\mathcal{P}_2$ are precisely

$$\{\{P^\phi \mid P \in \ell\} \mid \ell \in \mathcal{L}_1\}.$$

One similarly defines the notion of isomorphism and automorphism (or collineation) for affine planes.

There is an intimate relation between affine planes and projective planes. Namely, let $\mathcal{A}$ be any affine plane. Create a **slope point** (or **point at infinity**) for each parallel class of lines in $\mathcal{A}$, and add this slope point to each line of that parallel class. Then create a **line at infinity**, denoted $\ell_\infty$, which contains precisely the points at infinity, one for each parallel class of lines in $\mathcal{A}$. The resulting structure is a projective plane $\bar{\mathcal{A}}$, called the **projective completion** of the affine plane $\mathcal{A}$. Similarly, if $\mathcal{P}$ is any projective plane and $\ell$ is any line of $\mathcal{P}$, then removing the line $\ell$ and all its points from $\mathcal{P}$ will produce an affine plane, denoted by $\mathcal{P}^\ell$ or by $\mathcal{P} \setminus \ell$. Note that the lines of $\mathcal{P}^\ell$ will have one less point than the lines of $\mathcal{P}$. It is important to realize that the projective completion of an affine plane is uniquely determined, up to isomorphism, but it is possible to obtain nonisomorphic affine planes by deleting different lines from a given projective plane.

*Example 1.7.* The projective completion of $\mathbb{R}^2$ is a projective plane, called the **real projective plane**.

*Example 1.8.* Let $V$ be a three-dimensional vector space over some field (or skew field) $F$. Take as *points* the one-dimensional subspaces of $V$ and as *lines* the two-dimensional subspaces of $V$. A point is said to lie on a line if the one-dimensional subspace associated with the point is contained in the two-dimensional subspace associated with the line. Then the resulting structure is a projective plane, denoted by $\mathrm{PG}(2, F)$. Once again, if $F$ is a finite field, one obtains a finite projective plane. It is an interesting exercise to show that if $F$ is the field of real numbers, then the resulting projective plane is isomorphic to the real projective plane constructed in Example 1.7.

In checking that $\mathrm{PG}(2, F)$ is indeed a projective plane, one uses the **Dimension Theorem** from elementary linear algebra. Namely, if $A$ and $B$ are subspaces of any finite-dimensional vector space, then

$$\dim(A) + \dim(B) = \dim(A + B) + \dim(A \cap B).$$

In particular, if $V$ is a three-dimensional vector space, then any two distinct two-dimensional subspaces must meet in a uniquely determined one-dimensional subspace. That is, any two distinct lines of $\mathrm{PG}(2, F)$ meet in a unique point. The dimension theorem will play a significant role throughout this book.

In the projective plane $\mathrm{PG}(2, F)$ one naturally thinks of lines as having "dimension 1" and points as having "dimension 0". That is, the **projective dimension** of an object in $\mathrm{PG}(2, F)$ is one less than the vector space dimension of the associated vector subspace. This will hold true when we take higher dimensional analogues of the previous example.

It should be remarked at this stage that there do exist projective planes which are not isomorphic to $\mathrm{PG}(2, F)$ for any field (or skew field) $F$. One such example is the so-called Moulton plane, constructed by Moulton in 1902 (see [46], for instance). The Moulton plane is an infinite projective plane. It turns out that there are also (infinitely) many examples of finite projective planes which are not isomorphic to $\mathrm{PG}(2, F)$, some of which will be described later in this book. The interested reader should look at reference [49]. The projective planes isomorphic to some $\mathrm{PG}(2, F)$ will be called **classical projective**

**planes** or **field planes**. We use similar terminology for the affine planes iso-morphic to some $AG(2, F)$. In fact, one can see that removing any line and all its points from some $PG(2, F)$ will yield an isomorphic copy of $AG(2, F)$.

We now discuss in more detail the finite setting, which will be of most interest to us. In the classical case, necessarily $F = GF(q)$ for some prime power $q$, where $GF(q)$ denotes the Galois field of order $q$ (unique up to iso-morphism). In this case we denote $PG(2, F)$ and $AG(2, F)$ by $PG(2, q)$ and $AG(2, q)$, respectively, since there is no ambiguity about the field. Straight-forward counting shows that each line of $PG(2, q)$ has $q + 1$ points, each point lies on $q + 1$ lines, and the total number of points (or lines) is $q^2 + q + 1$. This pattern holds in general for any finite projective plane, whether or not it is classical. We omit the elementary counting argument for the following result.

**Theorem 1.9.** *Let $P$ be any finite projective plane. Then there is some integer $n \geq 2$, called the* **order** *of $P$, such that*

1. *each line of $P$ contains $n + 1$ points,*
2. *each point of $P$ lies on $n + 1$ lines,*
3. *the number of points of $P$ is $n^2 + n + 1$,*
4. *the number of lines of $P$ is $n^2 + n + 1$.*

Using the fact that $\bar{A}^{\ell_\infty} = A$ for any affine plane $A$, one immediately obtains the following result.

**Theorem 1.10.** *Let $A$ be any finite affine plane. Then there is some integer $n \geq 2$, called the* **order** *of $A$, such that*

1. *each line of $A$ contains $n$ points,*
2. *each point of $A$ lies on $n + 1$ lines,*
3. *the number of points of $A$ is $n^2$,*
4. *the number of lines of $A$ is $n^2 + n$.*

It should be remarked that all known finite projective (or affine) planes have prime power order. It is a very old and famous problem to decide if this must be true for every finite (projective or affine) plane. Currently, the smallest open case is $n = 12$. Of course, for every prime power $q$ there is at least one plane of order $q$, namely, the classical one associated with the finite field $GF(q)$. The smallest prime power for which there are nonclassical planes is $q = 9$. There are precisely four projective planes and seven affine planes of order 9 (it turns out that exactly two nonisomorphic affine planes are obtained from each nonclassical projective plane of order 9).

We conclude this section by considering higher dimensions. One can ax-iomatize a **projective geometry** in general (see [96], for instance), although we will not do so here. In this abstract setting one can define the notions of *subspace, span, independence*, and *dimension*. In so doing, the projective geome-tries of dimension 2 are precisely the projective planes as previously axiom-atized. The notions of isomorphism and automorphism (or collineation) also extend naturally to higher dimensions.

*Example 1.11.* Let $V$ be a $(d+1)$-dimensional vector space over some field (or skew field) $F$, for any integer $d \geq 2$. Take as *points* the one-dimensional subspaces of $V$, as *lines* the two-dimensional subspaces of $V$, as *planes* the three-dimensional subspaces of $V$, as *solids* the four-dimensional subspaces of $V$, ..., and as *hyperplanes* the $d$-dimensional subspaces of $V$. Then, using containment of subspaces, the resulting structure satisfies the axioms for a projective geometry of dimension $d$, called the **classical projective geometry of dimension** $d$ and denoted by $\mathrm{PG}(d, F)$. If $F$ is a finite field, so that $F = \mathrm{GF}(q)$ for some prime power $q$, then we obtain a finite projective geometry, denoted by $\mathrm{PG}(d, q)$.

Note that in this classical setting, the **projective dimension** of any subspace is one less than the dimension of the associated vector subspace. Hence the Dimension Theorem still remains valid. For the remainder of this book, when discussing classical projective geometries, the word **dimension** will always refer to projective dimension, unless specifically stated to the contrary.

An amazing result, a very nice development of which may be found in [46] or [85], is the following.

**Theorem 1.12.** *Let $\Pi$ be a projective geometry of dimension $d \geq 3$. Then $\Pi$ is isomorphic to $\mathrm{PG}(d, F)$ for some skew field $F$. In particular, if $\Pi$ is finite, then $\Pi$ is isomorphic to $\mathrm{PG}(d, q)$ for some prime power $q$, as every finite skew field is a field.*

Thus the only nonclassical projective geometries are the nonclassical projective planes alluded to above. Similarly, one can axiomatize affine geometries in general, and all of these are classical when the dimension is at least three. Removing any hyperplane and all subspaces contained in it from $\mathrm{PG}(d, F)$ will produce a **classical affine geometry of dimension** $d$, denoted by $\mathrm{AG}(d, F)$ or simply $\mathrm{AG}(d, q)$ when $F = \mathrm{GF}(q)$. As in the case $d = 2$, this classical affine geometry is isomorphic to the one obtained by taking a $d$-dimensional vector space over $F$, all its vectors as points, all cosets of one-dimensional subspaces as lines, all cosets of two-dimensional subspaces as planes, and so on.

It should be mentioned that classical projective geometries, including classical projective planes, are also called **Desarguesian**. This is because a result involving a certain configuration (called a *Desargues' configuration*) holds in a projective geometry if and only if the projective geometry arises from a vector space over a skew field; that is, if and only if the projective geometry is classical as previously defined (see [140]).

In the classical (Desarguesian) setting, one has the advantage of working with the underlying vector space. In particular, this allows one to naturally coordinatize. Consider the projective geometry $\mathrm{PG}(d, F)$ for some (skew) field $F$, and let $V$ denote the underlying $(d+1)$-dimensional vector space over $F$. Fix some ordered basis for $V$, and represent each vector uniquely as a $(d+1)$-tuple of scalars with respect to this basis. Since each point $P$ of $\mathrm{PG}(d, F)$ is a one-dimensional subspace $\langle v \rangle$ of $V$, the coordinates of $P$ are determined only up to nonzero scalar multiples. That is, any nonzero scalar multiple of $v$ will represent the same point $P$. These are the so-called **homogeneous coordinates** of $P$. Note that not all coordinates of $P$ can be 0 since $v$ is not the zero vector. That is, points of $PG(d, F)$ have homogeneous coordinates

$$\{(x_0, x_1, \ldots, x_d) \mid x_0, x_1, \ldots, x_d \in F, \text{ not all zero}\}$$

with the convention that $(x_0, x_1, \ldots, x_d)$ and $t(x_0, x_1, \ldots, x_d)$, $t \in F \setminus \{0\}$, represent the same point. One could obtain a unique representation for points by "left-normalizing"; that is, by making the first nonzero coordinate from the left equal to 1 via an appropriate scalar multiplication. Hence the points of $PG(2, F)$ are uniquely represented by

$$\{(0, 0, 1)\} \cup \{(0, 1, z) \mid z \in F\} \cup \{(1, y, z) \mid y, z \in F\}.$$

One similarly gets a unique representation for the points by "right-normalizing" the homogeneous coordinates.

Any hyperplane $H$ of $PG(d, F)$ is represented by a homogeneous linear equation

$$a_0 X_0 + a_1 X_1 + \cdots + a_d X_d = 0$$

in the $d + 1$ variables $X_0, X_1, \ldots, X_d$ for some coefficients $a_0, a_1, \ldots, a_d \in F$, not all 0. The ordered coefficients $(a_0, a_1, \ldots, a_d)$ of such an equation, unique up to nonzero scalar multiples, are the **homogeneous dual coordinates** of the hyperplane $H$. That is, $(a_0, a_1, \ldots, a_d)$ and $t(a_0, a_1, \ldots, a_d)$, $t \in F \setminus \{0\}$, represent the same hyperplane. One can make this representation unique by (left or right) normalizing as above. By convention we will think of point coordinates as row vectors and hyperplane coordinates as column vectors. Note that a point $P = (x_0, x_1, \ldots, x_d)$ lies on a hyperplane $H = (a_0, a_1, \ldots, a_d)'$, where $'$ denotes the transpose of the vector, if and only if the dot product $(x_0, x_1, \ldots, x_d)(a_0, a_1, \ldots, a_d)'$ equals 0. This is a very handy computational tool.

*Example 1.13.* Recall the real projective plane; namely, the projective completion of $\mathbb{R}^2$. If we think of the line at infinity as having equation $Z = 0$, then an affine point has homogeneous coordinates $(x, y, z)$, for $x, y, z \in \mathbb{R}$ with $z \neq 0$ (unique up to nonzero scalar multiplication). The cartesian coordinates for the corresponding point of $\mathbb{R}^2$ are $(x/z, y/z)$. A line of $\mathbb{R}^2$ with cartesian equation $aX + bY + c = 0$ has a corresponding homogeneous equation $aX + bY + cZ = 0$ with homogeneous coordinates $(a, b, c)'$. The point at infinity of this line is $(b, -a, 0)$. In general, points at infinity have homogeneous coordinates $(x, y, 0)$, for $x, y \in \mathbb{R}$ (unique up to nonzero scalar multiplication). The line at infinity has homogeneous coordinates $\ell_\infty = (0, 0, 1)'$.

**Definition 1.14.** *Let $V$ be a vector space over the field $F$, and let $\alpha$ be an automorphism of the field $F$. A mapping $T : V \to V$ is called a* **semilinear transformation** *with* **companion automorphism** $\alpha$ *provided*

$$T(v + w) = T(v) + T(w), \ \forall \ v, w \in V, \text{ and}$$
$$T(\lambda v) = \lambda^\alpha T(v), \ \forall \ v \in V, \ \forall \ \lambda \in F.$$

*A semilinear transformation is called* **nonsingular** *if it is a bijection. If $\alpha$ is the identity automorphism, then $T$ is called a* **linear transformation**.

In the classical setting any nonsingular semilinear transformation of the underlying vector space is a bijection mapping subspaces to subspaces and thus induces an automorphism (collineation) of the classical projective geometry. Incredibly, these are the only automorphisms in this case. A proof of the following fundamental result may be found in [46] (or in [140] for classical projective planes).

**Theorem 1.15 (Fundamental Theorem of Projective Geometry).** *Every automorphism (collineation) of $\mathrm{PG}(d, F)$, $d \geq 2$, is induced by a nonsingular semilinear transformation of the underlying vector space.*

The set of all automorphisms of $\mathrm{PG}(d, F)$ is a group under composition of maps, denoted by $\mathrm{P\Gamma L}(d + 1, F)$. It is the quotient group of $\mathrm{\Gamma L}(d + 1, F)$ by its center $Z$, where $\mathrm{\Gamma L}(d + 1, F)$ is the group of all nonsingular semilinear transformations of a $(d + 1)$-dimensional vector space over $F$. The center $Z$ is the subgroup of all nonzero scalar multiples of the identity transformation. The (normal) subgroup of $\mathrm{P\Gamma L}(d + 1, F)$ consisting of all automorphisms induced by nonsingular linear transformations of the underlying vector space is denoted by $\mathrm{PGL}(d + 1, F)$, and is called the **homography subgroup** or the **subgroup of projectivities**. It is the quotient group of the general linear group $\mathrm{GL}(d + 1, F)$ by $Z$. Hence a homography $\tau$ can be represented by a nonsingular "homogeneous" matrix $A$. That is, if $v$ is the row vector of homogeneous coordinates of some point $P$, then $vA$ is the row vector of homogeneous coordinates of the image point $P^\tau$. The automorphisms in $\mathrm{P\Gamma L}(d + 1, F) \setminus \mathrm{PGL}(d + 1, F)$ are typically called **nonlinear collineations**. Moreover, the nonlinear collineations induced solely by a field automorphism are called **automorphic collineations**. If $P = (x_0, x_1, \ldots, x_d)$ is a point of $\mathrm{PG}(d, F)$ and $\alpha$ is an automorphism of $F$, then we use the notation $P^\alpha$ for the image of $P$ under the resulting automorphic collineation, where $P^\alpha = (x_0^\alpha, x_1^\alpha, \ldots, x_d^\alpha)$.

The Fundamental Theorem of Projective Geometry implies the following:

*every collineation of a classical projective geometry is the product of a homography and an automorphic collineation.*

## 1.2 Finite Fields

In this section we review a few basic ideas concerning finite fields. Recall that every finite field must have prime power order, and given any prime power $q$, there is a unique field (up to isomorphism) with $q$ elements. As introduced in the previous section, we will use the notation $\mathrm{GF}(q)$ to denote this field. The multiplicative group $\mathrm{GF}(q) \backslash \{0\}$ of $\mathrm{GF}(q)$ is cyclic, and any generator of this multiplicative group is called a **primitive element** of the field. The additive group of $\mathrm{GF}(q)$ is an elementary abelian $p$-group, where $q$ is a power of the prime $p$. That is, adding any element of $\mathrm{GF}(q)$ to itself $p$ times will yield 0, and the field has **characteristic** $p$. We denote this by writing $\mathrm{char}(\mathrm{GF}(q)) = p$. Hence, using the Binomial Theorem, we see that $(x+y)^p = x^p + y^p$ for any $x, y \in \mathrm{GF}(q)$. Therefore the mapping $x \mapsto x^p$ is a field automorphism of $\mathrm{GF}(q)$, often called the **Frobenius automorphism**. In fact, if $q = p^e$, the automorphism group of $\mathrm{GF}(q)$ is cyclic of order $e$, generated by the Frobenius automorphism.

If $q$ is even (that is, $q = 2^e$ for some positive integer $e$), then $x \mapsto x^2$ is a field automorphism and every element of $\mathrm{GF}(q)$ is a square. For $q$ even, the quadratic equation $aX^2 + bX + c = 0$, with $a, b, c \in \mathrm{GF}(q)$ and $a \neq 0$, is **irreducible** over $\mathrm{GF}(q)$ (that is, has no roots in $\mathrm{GF}(q)$) if and only if $b \neq 0$ and $T(ac/b^2) = 1$, where $T : \mathrm{GF}(2^e) \to \mathrm{GF}(2)$ via $x \mapsto x + x^2 + x^4 + \cdots + x^{2^{e-1}}$ is the **absolute trace** function. The above quadratic equation has a unique root in $\mathrm{GF}(2^e)$ if and only if $b = 0$. It is useful to note that the absolute trace satisfies the properties $T(x + y) = T(x) + T(y)$ and $T(x^2) = T(x)$ for all $x, y \in \mathrm{GF}(2^e)$. In particular, $T(x^2 + x) = 0$ for all $x \in \mathrm{GF}(2^e)$.

If $q$ is odd, then precisely half the nonzero elements are squares and half the nonzero elements are nonsquares. For odd $q$, the quadratic equation $aX^2 + bX + c = 0$, with $a, b, c \in \mathrm{GF}(q)$ and $a \neq 0$, is irreducible if and only if the **discriminant** $b^2 - 4ac$ is a nonsquare in $\mathrm{GF}(q)$. The equation has a unique root in $\mathrm{GF}(q)$ if and only if $b^2 - 4ac = 0$. So, for odd $q$, quadratic equations over $\mathrm{GF}(q)$ act much like quadratic equations over the real numbers $\mathbb{R}$, and the usual quadratic formula for finding the roots of a quadratic equation still holds.

In this book we will primarily be interested in finite fields which admit an **involutory** field automorphism; that is, an automorphism of order 2. Hence the order of such a field must be square, and we will typically redefine $q$ so that the field in question is $\mathrm{GF}(q^2)$. The mapping

$$\sigma : x \longmapsto x^q$$

will then be the (unique) involutory field automorphism of $\mathrm{GF}(q^2)$. Its fixed field is the subfield $\mathrm{GF}(q)$ of $\mathrm{GF}(q^2)$. The automorphism $\sigma$ is often called **conjugation**, as it is analogous to complex conjugation on the field of complex numbers with the real numbers as the fixed field.

If $x \in \mathrm{GF}(q^2)$, then $x^{q+1} \in \mathrm{GF}(q)$ and the mapping $x \mapsto x^{q+1}$ is called the **norm** function from $\mathrm{GF}(q^2)$ to $\mathrm{GF}(q)$. The only element of norm 0 in $\mathrm{GF}(q^2)$ is

0. If $a$ is some nonzero element of $GF(q)$, then a polynomial argument shows that at most $q + 1$ elements of $GF(q^2)$ have norm $a$. Since there are $q^2 - 1$ nonzero elements of $GF(q^2)$, we see that there are precisely $q + 1$ elements of norm $a$ in $GF(q^2)$ for each nonzero element $a$ of the subfield $GF(q)$. Similarly, the element $x + x^q$ is in $GF(q)$ for each $x \in GF(q^2)$, and is called the **trace** of $x$ over $GF(q)$. Each element of $GF(q)$, including 0, has $q$ pre-images in $GF(q^2)$ under the trace from $GF(q^2)$ to $GF(q)$.

## 1.3 Quadrics in Low Dimensions

In [131] a thorough discussion of quadrics in finite projective geometries is given, including the development of canonical forms in all cases. Here we briefly discuss, without proofs, the quadrics that will be of importance to us.

**Definition 1.16.** *A* **quadric** *in* $PG(d, F)$ *is a set consisting of all the points whose homogeneous coordinates satisfy some nonzero homogeneous equation of degree 2. If the equation can be reduced to fewer than $d + 1$ variables by a change of basis, we call the quadric* **singular** *or* **degenerate**. *Otherwise, we call the quadric* **nonsingular** *or* **nondegenerate**. *If $d = 2$, a quadric is called a* **conic**. *A nondegenerate conic is also called an* **irreducible conic**.

If the field $F$ is infinite, quadrics may be empty. However, this cannot happen over finite fields. In fact, quadrics over finite fields are quite restricted. For the remainder of this section we assume $F$ is finite.

When $d = 2$, an irreducible conic in $PG(2, q)$ is uniquely determined, up to a change of basis. A convenient canonical form for an irreducible conic $C$ is

$$X_1^2 - X_0 X_2 = 0,$$

so that using left-normalized coordinates we have

$$C = \{(0, 0, 1)\} \cup \{(1, y, y^2) \mid y \in GF(q)\}.$$

In particular, $C$ has $q + 1$ points with no three collinear. Thus there is a unique tangent line at each point of $C$. It should be noted that any five points of $PG(2, q)$, with no three collinear, uniquely determine an irreducible conic. In particular, any irreducible conic is determined by any five of its points.

Straightforward computations show that for odd $q$, every point of $PG(2, q)$ not on $C$ lies on 0 or 2 tangents, the so-called **interior** and **exterior** points of $C$, respectively. Thus for odd $q$ an irreducible conic in $PG(2, q)$ "looks" like an ellipse in the Euclidean plane $\mathbb{R}^2$. However, for even $q$ the $q + 1$ tangents to $C$ all pass through the common point $N = (0, 1, 0)$, called the **nucleus** of $C$. Thus for even $q$, irreducible conics in $PG(2, q)$ and ellipses in $\mathbb{R}^2$ are quite different objects.

There are three types of degenerate conics in $PG(2, q)$: a point, a line, and a pair of intersecting lines. Canonical forms for these degenerate conics are

$$X_0^2 + bX_0X_1 + cX_1^2 = 0, \quad X_0^2 = 0, \quad \text{and} \quad X_0X_1 = 0,$$

respectively, where $X_0^2 + bX_0X_1 + cX_1^2$ is an **irreducible binary quadratic form** (that is, $X^2 + bX + c = 0$ is an irreducible quadratic equation over $GF(q)$ as previously defined).

When $d = 3$, there are two types of nonsingular quadrics in $PG(3, q)$: the **elliptic quadric** and the **hyperbolic quadric**. The elliptic quadric consists of $q^2 + 1$ points with no three collinear. There is a unique **tangent plane** at each point of the quadric, which meets the elliptic quadric only in that point, and all other planes of $PG(3, q)$ meet the quadric in an irreducible conic. The latter planes are called **secant planes** of the elliptic quadric. A canonical form for the elliptic quadric in $PG(3, q)$ is

$$f(X_0, X_1) + X_2X_3 = 0,$$

where $f$ is an irreducible binary quadratic form. The hyperbolic quadric of $PG(3, q)$ consists of $(q + 1)^2$ points and is *ruled* by two different families of $q + 1$ mutually skew lines, with any line from one family meeting every line of the other family in a unique point of the quadric. Every plane of $PG(3, q)$ meets the hyperbolic quadric in an irreducible conic or a pair of intersecting lines, one from each of the above families. The latter planes are called **tangent planes** of the hyperbolic quadric. A canonical form for the hyperbolic quadric in $PG(3, q)$ is

$$X_0X_1 + X_2X_3 = 0.$$

There are four types of degenerate quadrics in $PG(3, q)$: a line, a plane, a pair of intersecting planes (necessarily meeting in a line by the Dimension Theorem), and a **quadratic cone**. By a quadratic cone, we mean a three-dimensional cone whose base is an irreducible conic and whose vertex is a point not lying in the plane of the base. As is true for all cones, the points of the cone are the points lying on the lines joining the vertex to a point of the base. This cone is also called a **conic cone**. Canonical forms for these four degenerate quadrics are

$$f(X_0, X_1) = 0, \quad X_0^2 = 0, \quad X_0X_1 = 0, \quad \text{and} \quad X_1^2 - X_0X_2 = 0,$$

respectively, where $f$ is an irreducible binary quadratic form.

It should be mentioned that cones can be defined much more generally, and they will appear throughout the book. We will mostly be interested in **quadric cones**. That is, suppose $\Pi_1$ and $\Pi_2$ are disjoint subspaces in some projective geometry $PG(d, F)$, and suppose that $\mathcal{Q}$ is some nonsingular quadric in $\Pi_2$. The set consisting of all points lying on the lines joining a point of $\Pi_1$ and a point of $\mathcal{Q}$ is called the **quadric cone** with **vertex** $\Pi_1$ and

**base** $Q$. If $\Pi_1$ is a single point, sometimes the cone is called the **point cone** over $Q$.

When $d = 4$, there is a unique nonsingular quadric and six degenerate quadrics in $PG(4, q)$, up to a change of basis. The nonsingular quadric is a **parabolic quadric**. It consists of $q^3 + q^2 + q + 1$ points and has $q^3 + q^2 + q + 1$ lines lying on it, with $q + 1$ such lines through each of its points. A canonical form for this nonsingular quadric is

$$X_0^2 + X_1 X_2 + X_3 X_4 = 0.$$

Of the six degenerate quadrics in $PG(4, q)$, we only will be interested in one of them, the **elliptic cone** (or **orthogonal cone**). This is a four-dimensional cone whose base is a three-dimensional elliptic quadric and whose vertex is a point not contained in the hyperplane of the base. A canonical form for this cone is

$$f(X_0, X_1) + X_2 X_3 = 0,$$

where $f$ is some irreducible binary quadratic form.

Since later on we will be extensively using this four-dimensional elliptic cone, we develop a few useful facts at this time. Let $Q$ be such an elliptic cone in $PG(4, q)$, with vertex $P$ and base $\mathcal{E}$, where $\Sigma \cong PG(3, q)$ is the hyperplane (solid) containing the base. The lines joining $P$ to the $q^2 + 1$ points of $\mathcal{E}$ are the **generators** of $Q$, and these are the only lines contained in $Q$. As no three points of $\mathcal{E}$ are collinear, no three generators of $Q$ are coplanar. Hence there are three choices for the intersection of $Q$ with any plane through its vertex $P$: namely, the vertex $P$ itself, some generator of $Q$, or a pair of generators of $Q$. A plane meeting $Q$ in some generator $g$ will be called a **tangent plane** at $g$. Let $\Sigma' \cong PG(3, q)$ be any hyperplane containing the vertex $P$. Then $\Sigma' \cap \Sigma$ is a plane by the Dimension Theorem, and this plane meets $\mathcal{E}$ either in a point or an irreducible conic as previously discussed. Thus $\Sigma'$ meets $Q$ either in a single generator or a quadratic cone of generators. This further implies that a plane $\pi$ not through the vertex $P$ will meet $Q$ in a single point or an irreducible conic; namely, look at the intersection of $\Sigma' = \langle \pi, P \rangle$ with $Q$. A hyperplane meeting $Q$ in a generator $g$ will be called a **tangent hyperplane** at $g$. Of course, any hyperplane not through the vertex $P$ meets $Q$ in a three-dimensional elliptic quadric.

**Theorem 1.17.** *Let $Q$ be an elliptic cone of $PG(4, q)$, and let $g$ be some generator of $Q$. Then there is a unique tangent hyperplane at $g$. Moreover, all the tangent planes at $g$ are contained in this unique tangent hyperplane.*

*Proof.* Any hyperplane through $g$ meets $Q$ in $g$ or in a quadratic cone containing $g$. Since no three generators of $Q$ are coplanar, any three distinct generators determine a hyperplane that meets $Q$ in a quadratic cone. As there are $q^2$ generators of $Q$ other than $g$, we see that the number of distinct hyperplanes through $g$ meeting $Q$ in a quadratic cone is

$$\binom{q^2}{2} \Big/ \binom{q}{2} = q^2 + q.$$

But the number of hyperplanes of $PG(4, q)$ through $g$ is $q^2 + q + 1$ (the same number as the number of points on a plane of $PG(4, q)$, by duality), and hence there must be exactly one tangent hyperplane at $g$.

Since no three generators of $Q$ are coplanar, the number of distinct planes through $g$ meeting $Q$ in a pair of generators is $q^2$. As there are $q^2 + q + 1$ planes of $PG(4, q)$ through $g$ (the same number as the number of lines in a plane of $PG(4, q)$, by duality), we see that there are exactly $q + 1$ tangent planes through $g$. Since the unique tangent hyperplane at $g$ is a hyperplane, it has exactly $q + 1$ planes through $g$ (similar counting), all of which necessarily meet $Q$ precisely in $g$. Thus all tangent planes at $g$ are contained in the unique tangent hyperplane at $g$.                                   □

## 1.4 Ovals and Ovoids

In this section we give combinatorial generalizations of the conic, the elliptic quadric, and the elliptic cone in various finite projective spaces.

**Definition 1.18.** *An **oval** in a projective plane of order $q$ is a set of $q + 1$ points, no three collinear. An **ovoid** in $PG(3, q)$, $q > 2$, is a set of $q^2 + 1$ points, no three collinear. An **ovoidal cone** in $PG(4, q)$ is a cone whose vertex is a point $P$ and whose base is an ovoid in a hyperplane not containing $P$.*

The classical example of an oval in $PG(2, q)$ is an irreducible conic. A remarkable result of Segre [195] shows that for odd $q$, every oval in $PG(2, q)$ is a conic. If $q$ is even, there are many examples of ovals in $PG(2, q)$ that are not conics.

The classical example of an ovoid in $PG(3, q)$ is an elliptic quadric. Moreover, if $q$ is odd, all ovoids are elliptic quadrics (see [28] or [176]). However, in even characteristic there is one known family of ovoids that are not elliptic quadrics. These were discovered by Tits [212], and are now called the **Suzuki-Tits ovoids** since the Suzuki group naturally acts on these ovoids. They exist precisely when $q \geq 8$ is an odd power of 2. A very famous open problem is to determine if there are any other ovoids in $PG(3, q)$. In any case, the combinatorics of planar intersections for ovoids in $PG(3, q)$ is exactly the same as for elliptic quadrics (see [130]). Namely, at each point of an ovoid there is a unique tangent plane, and all other planes meet the ovoid in a set of $q + 1$ points, no three of which are collinear. That is, planes meet an ovoid in a point or an oval.

The classical example of an ovoidal cone in $PG(4, q)$ is an elliptic cone. These are the only ones for odd $q$. However, other ovoidal cones can be constructed by taking a Suzuki-Tits ovoid as a base for the cone when $q \geq 8$ is an odd power of 2.

In the previous section we discussed the possible intersections for a plane and a hyperplane with an elliptic cone in $PG(4, q)$. Indeed, the exact same possibilities occur for an ovoidal cone, for exactly the same reasons. That is, any plane in $PG(4, q)$ meets an ovoidal cone $\mathcal{U}$ in a point, a line, a pair of intersecting lines, or an oval. In particular, a plane meets $\mathcal{U}$ in 1, $q + 1$, or $2q + 1$ points. Similarly, any hyperplane of $PG(4, q)$ meets an ovoidal cone $\mathcal{U}$ in a line, a point cone over an oval, or an ovoid. As with elliptic cones, if $P$ is the vertex of the ovoidal cone $\mathcal{U}$, then the **generators** of $\mathcal{U}$ are the lines joining $P$ to a point of the base ovoid. These are the only lines lying completely on $\mathcal{U}$. A hyperplane meeting $\mathcal{U}$ precisely in a generator $g$ is called a **tangent hyperplane** at $g$, and a plane meeting $\mathcal{U}$ in a generator $g$ is called a **tangent plane** at $g$.

**Theorem 1.19.** *Let $\mathcal{U}$ be an ovoidal cone in $PG(4, q)$ with vertex $P$, and let $g$ be some generator of $\mathcal{U}$. Then the following hold.*

1. *There is a unique tangent hyperplane to $\mathcal{U}$ at $g$. All tangent planes to $\mathcal{U}$ at $g$ are contained in the unique tangent hyperplane at $g$.*
2. *Any plane of $PG(4, q)$ which meets $\mathcal{U}$ precisely in a point of $g \backslash P$ is necessarily contained in the unique tangent hyperplane at $g$.*

*Proof.* The proof of (1) is a counting argument that follows exactly as in the proof of Theorem 1.17. Then (2) follows immediately from (1), since any such plane together with a generator $g$ determines a hyperplane which must meet $\mathcal{U}$ only in $g$. □

## 1.5 Some Linear Algebra

The final piece of preparation we need is a little linear algebra. Although we could develop the needed theory for (left or right) vector spaces over skew fields, this added abstraction would not prove to be useful for our purposes. Thus we restrict ourselves to vector spaces over fields.

**Definition 1.20.** *Let $\alpha$ be some automorphism of the field $F$, and let $V$ be a vector space over $F$. Then a **sesquilinear form** with **companion automorphism** $\alpha$ is a mapping $s : V \times V \longmapsto F$ such that*

1. *$s(v_1 + v_2, w_1 + w_2) = s(v_1, w_1) + s(v_1, w_2) + s(v_2, w_1) + s(v_2, w_2)$ for all $v_1, v_2, w_1, w_2 \in V$,*
2. *$s(\lambda v, w) = \lambda s(v, w)$ for all $\lambda \in F$ and for all $v, w \in V$,*
3. *$s(v, \lambda w) = \lambda^\alpha s(v, w)$ for all $\lambda \in F$ and for all $v, w \in V$.*

*Furthermore, we say that $s$ is **nondegenerate** provided the only vector $v$ in $V$ with $s(v, w) = 0$ for all $w \in V$ is $v = 0$.*

It is a useful exercise to show that one can equivalently define nondegeneracy of $s$ by interchanging the roles of $v$ and $w$ in the above definition.

In general, if $W$ is any subspace of $V$, we define

$$W^\perp = \{v \in V \mid s(v, w) = 0 \ \forall \ w \in W\}.$$

We think of $W^\perp$ as being the **orthogonal complement** of the subspace $W$ with respect to the sesquilinear form $s$. In the finite-dimensional case we would like to show that the usual dimension result for orthogonal complements (when working with dot products over $\mathbb{R}^n$) holds in this more abstract setting. To do this, we use the dual space of $V$.

Assume now that $V$ is finite-dimensional, and let $V^*$ be the **dual space** of $V$. That is, $V^*$ is the vector space of all linear functionals on $V$. Using the standard dual basis, we see that $\dim(V^*) = \dim(V)$. If $W$ is any subspace of $V$, we let

$$W^\circ = \{f \in V^* \mid f(w) = 0 \ \forall \ w \in W\}$$

be the **annihilator** of $W$. Similarly, if $A$ is a subspace of $V^*$, we let

$$A^\circ = \{v \in V \mid f(v) = 0 \ \forall \ f \in A\}$$

be the annihilator of $A$.

**Theorem 1.21.** *Let $V$ be a finite-dimensional vector space. Let $W$ be a subspace of $V$, and let $A$ be a subspace of $V^*$. Then*

$$\dim(V) = \dim(W) + \dim(W^\circ) = \dim(A) + \dim(A^\circ).$$

*Proof.* By symmetry it suffices to prove that $\dim(V) = \dim(W) + \dim(W^\circ)$. Let $\phi : V^* \mapsto W^*$ be the restriction map. Then $\phi$ is a homomorphism on the additive group $V^*$. Moreover, $\ker(\phi) = W^\circ$. Since $\phi$ is surjective, we have $V^*/W^\circ \cong W^*$ by the First Homomorphism Theorem. Hence

$$\dim(W^*) = \dim(V^*) - \dim(W^\circ).$$

Since $\dim(V^*) = \dim(V)$ and $\dim(W^*) = \dim(W)$, the result follows. $\square$

**Theorem 1.22.** *Let $V$ be a finite-dimensional vector space over a field $F$, equipped with a nondegenerate sesquilinear form $s$. Let $W$ be a subspace of $V$, and let $W^\perp$ be the orthogonal complement of $W$ with respect to $s$. Then*

$$\dim(W^\perp) = \dim(V) - \dim(W).$$

*Proof.* For each $v \in V$, let $f_v : V \mapsto F$ via $f_v(u) = s(u, v)$. Since $s$ is linear in the first component, $f_v$ is a linear functional on $V$. Let $A_W = \{f_v \mid v \in W\}$. Straightforward computations show that $A_W$ is a subspace of $V^*$ under the usual operations. Moreover,

$$\begin{aligned}
W^\perp &= \{v \in V \mid s(v, w) = 0 \ \forall \ w \in W\} \\
&= \{v \in V \mid f_w(v) = 0 \ \forall \ w \in W\} \\
&= \{v \in V \mid f(v) = 0 \ \forall \ f \in A_W\} \\
&= A_W^\circ.
\end{aligned}$$

Hence by Theorem 1.21 we have

$$\dim(W^\perp) = \dim(A_W^\circ) = \dim(V) - \dim(A_W).$$

But the surjective semilinear map $W \mapsto A_W$ via $w \mapsto f_W$ is nonsingular since $s$ is nondegenerate, using the alternate (but equivalent) definition of nondegeneracy. Therefore $\dim(W) = \dim(A_W)$ and the result follows. $\square$

Thus the mapping $W \mapsto W^\perp$ on subspaces of $V$ sends $k$-dimensional vector subspaces to $(d+1-k)$-dimensional vector subspaces if $\dim(V) = d+1$. Clearly this mapping reverses containment, so that if $U \subseteq W$, then $W^\perp \subseteq U^\perp$. Moreover, from Theorem 1.22 it is straightforward to show that this mapping is a bijection on the subspaces of $V$. Hence, passing to the associated classical projective geometry $\mathrm{PG}(d, F)$, we get a bijection on the projective subspaces that reverses containment, and thus interchanges points and hyperplanes. It should be noted that this mapping does not necessarily have order two.

**Definition 1.23.** *A **correlation** of a projective geometry is a bijection on its subspaces that reverses containment. In particular, a correlation interchanges points and hyperplanes.*

The following result, first proved by Birkhoff and von Neumann [54] in 1936, shows the intimate connection between correlations and sesquilinear forms in the classical setting. It is analogous to the Fundamental Theorem of Projective Geometry, with collineations replaced by correlations. We omit the proof.

**Theorem 1.24.** *Let $\mathrm{PG}(d, F)$, $d \geq 2$, be a classical projective geometry over some field $F$ with underlying $(d+1)$-dimensional vector space $V$. If $\rho$ is any correlation of $\mathrm{PG}(d, F)$, then there exists some nondegenerate sesquilinear form $s$ on $V$ with an accompanying field automorphism which induces $\rho$. That is, $W^\rho = W^\perp$ for every subspace $W$, where $W^\perp$ is the orthogonal complement of $W$ with respect to $s$.*

If we really want to think of $s(v, w) = 0$ as meaning $v$ is "orthogonal" to $w$, then we want our sesquilinear form to have the property that

$$s(v, w) = 0 \Rightarrow s(w, v) = 0.$$

**Definition 1.25.** *A sesquilinear form on a vector space $V$ is called **reflexive** if for all $v, w \in V$, $s(v, w) = 0 \Rightarrow s(w, v) = 0$.*

**Theorem 1.26.** *Let $\rho$ be a correlation of the classical projective geometry $\mathrm{PG}(d, F)$, $d \geq 2$, with associated nondegenerate sesquilinear form $s$ on the underlying vector space $V$. Then $s$ is reflexive if and only if $\rho$ has order two.*

*Proof.* Assume first that $s$ is reflexive, and let $W$ be any subspace of $V$. Then $W^\perp = \{v \in V \mid s(v, w) = 0 \ \forall \ w \in W\}$, and hence

$$W^{\perp\perp} = \{v \in V \mid s(v, u) = 0 \; \forall \; u \in W^{\perp}\}.$$

But $w \in W$ implies $s(v, w) = 0$ for all $v \in W^{\perp}$, which in turn implies that $s(w, v) = 0$ for all $v \in W^{\perp}$ since $s$ is reflexive. That is, every vector $w$ of $W$ is in $W^{\perp\perp}$ and $W \subseteq W^{\perp\perp}$. Since $\dim(W^{\perp\perp}) = \dim(V) - \dim(W^{\perp}) = \dim(W)$, we have $W = W^{\perp\perp}$ and the correlation $\rho$ has order two.

Conversely, assume that $\rho$ has order two and thus $W = W^{\perp\perp}$ for all subspaces $W$ of $V$. Consider any two vectors $v$ and $w$ with $s(v, w) = 0$. Then $v \in \langle w \rangle^{\perp}$ and hence $\langle v \rangle \subseteq \langle w \rangle^{\perp}$. Therefore $\langle w \rangle = \langle w \rangle^{\perp\perp} \subseteq \langle v \rangle^{\perp}$ by our assumption with $W = \langle w \rangle$, and thus $s(w, v) = 0$. That is, $s$ is reflexive by definition. □

Therefore the correlations of order two are the natural ones to work with when doing geometry, since then our notion of "orthogonality" makes sense. If $v$ is "orthogonal" to $w$, we certainly want $w$ to be "orthogonal" to $v$.

**Definition 1.27.** *A correlation of order two is called a* **polarity**.

The following result, also proved by Birkhoff and von Neumann [54], gives all the possibilities for polarities of classical projective geometries. Again, we omit the proof.

**Theorem 1.28.** *Let $\rho$ be a polarity of the classical projective geometry* $\mathrm{PG}(d, F)$, *$d \geq 2$, and let $s$ be the associated nondegenerate and reflexive sesquilinear form on the underlying vector space $V$. Let $\alpha$ be the companion field automorphism for $s$. Then precisely one of the following occurs:*

1. *$\alpha$ is the identity and $s(v, w) = s(w, v)$ for all $v, w \in V$. If the characteristic of $F$ is 2, then $s(v, v) \neq 0$ for some $v \in V$.*
2. *$\alpha$ is the identity and $s(v, v) = 0$ for all $v \in V$.*
3. *$\alpha$ has order 2 and $s(v, w) = s(w, v)^{\alpha}$ for all $v, w \in V$.*

The polarities in the above list are called **orthogonal**, **symplectic**, and **unitary**, respectively, and the associated sesquilinear forms are called **symmetric bilinear**, **skew-symmetric bilinear**, and **Hermitian**, respectively. The symplectic polarity is sometimes called a **null** polarity since every vector is orthogonal to itself, and the associated sesquilinear form is sometimes called **alternating**. The polarity names come from the associated classical groups. When the characteristic of the field $F$ is 2, orthogonal polarities are typically called **pseudopolarities** for reasons that will soon become apparent. If the characteristic of $F$ is not 2, orthogonal polarities are usually called **ordinary** polarities.

If one fixes a basis for the underlying vector space $V$, say $\{b_0, b_1, \ldots, b_d\}$, and constructs the $(d + 1) \times (d + 1)$ matrix $G$ whose $(i, j)$-entry is $s(b_i, b_j)$, then using coordinates with respect to this fixed basis one computes $s(v, w)$ as the matrix product $xG(y^{\alpha})'$. Here $x$ and $y$ are the row vectors of coordinates for $v$ and $w$, respectively, with respect to the given basis, and the exponent $\prime$ denotes transpose as previously defined. The exponent $\alpha$ in the above expression means that the field automorphism $\alpha$ is applied to each component of the (row or column) vector. The matrix $G$ is often called the **Gram matrix** of $s$ with respect to the given basis. Since the sesquilinear form is nondegenerate, the Gram matrix is necessarily nonsingular. Note that $G$ is a symmetric matrix for orthogonal polarities; if the characteristic of the field is 2, then at least one diagonal entry of this symmetric matrix is nonzero. For symplectic polarities the Gram matrix is skew-symmetric (that is, $G' = -G$) with all diagonal entries equal to 0 in any characteristic. As nonsingular, skew-symmetric $(d + 1) \times (d + 1)$ matrices only exist when $d + 1$ is even, the projective dimension $d$ must be odd for symplectic polarities to exist. For unitary polarities the Gram matrix is Hermitian; that is, $G' = G^\alpha$ where $G^\alpha$ means $\alpha$ is applied to each entry of $G$. In particular, the diagonal entries of a Hermitian matrix must come from the subfield fixed by $\alpha$. Thus the names given above for the various sesquilinear forms arise from the classical names for the associated Gram matrices.

As discussed in Section 1.2, a finite field has an automorphism of order two if and only if it has square order. Thus unitary polarities exist for finite classical projective geometries if and only if $F = \mathrm{GF}(q^2)$ for some prime power $q$, in which case the involutory field automorphism $\alpha$ is the conjugation map $\sigma : x \mapsto x^q$ as previously defined. In this case we often write the Gram matrix product as $xG(y^q)'$, where the exponent $q$ now means that each entry of the (row or column) vector is raised to the $q^{\text{th}}$ power. Also in this case, as $G$ is Hermitian, the diagonal entries of $G$ are fixed by $\sigma$ and hence are from the subfield $\mathrm{GF}(q)$, while the off-diagonal entry $g_{ij}$ is necessarily equal to $g_{ji}^q$.

If $P$ is a point of $\mathrm{PG}(d, F)$, the hyperplane $P^\rho$ is called the **polar hyperplane** of $P$. Similarly, if $H$ is a hyperplane of $\mathrm{PG}(d, F)$, the point $H^\rho$ is called the **pole** of $H$. Using the associated sesquilinear form and its Gram matrix $G$ with respect to some fixed basis, one can easily compute poles and polars. Let $x$ be the row vector of homogeneous coordinates for $P$ with respect to the given basis, and let $y'$ be the column vector of homogeneous dual coordinates for $H$ with respect to the same basis. Then the polar of $P$ has dual homogeneous coordinates $G(x^\alpha)'$, and the pole of $H$ has homogeneous coordinates $y^\alpha G^{-1}$.

**Definition 1.29.** *Let $\rho$ be a polarity of some projective geometry $\Pi$. A subspace $A$ of $\Pi$ is called **totally isotropic** if $A \subseteq A^\rho$; **isotropic** if $A \cap A^\rho \neq \emptyset$; or **nonisotropic** if $A \cap A^\rho = \emptyset$. A point $P$ of $\Pi$ is called **absolute** if $P \in P^\rho$; else, $P$ is called **nonabsolute**. Similarly, a hyperplane $H$ of $\Pi$ is called **absolute** if $H^\rho \in H$; or **nonabsolute** if $H^\rho \notin H$.*

For classical projective geometries $\Pi = \mathrm{PG}(d, F)$, every point and hence every hyperplane of a symplectic (null) polarity is absolute. If the characteristic of the field $F$ is not 2, the nonsingular quadrics of $\mathrm{PG}(d, F)$ are precisely the sets of absolute points of orthogonal polarities. This is the main reason for calling these polarities *ordinary*, as mentioned above. For finite fields of characteristic 2, the absolute points of orthogonal polarities are easily described.

**Theorem 1.30.** *Let $\rho$ be an orthogonal polarity of* $\mathrm{PG}(d, 2^e)$. *Then the set of absolute points of $\rho$ is some hyperplane of* $\mathrm{PG}(d, 2^e)$.

*Proof.* Let $s$ be the associated symmetric bilinear form of $\rho$, and let $G$ be the Gram matrix of $s$ with respect to some fixed basis of the underlying vector space. Since the characteristic of the field is 2, at least one diagonal entry of the symmetric matrix $G$ is nonzero as previously discussed. Then a point $P$ with homogeneous coordinates $x$ with respect to this fixed basis is absolute if and only if $xGx' = 0$. Expanding this matrix equation and using the fact that the characteristic of the field is 2, we obtain the equation

$$g_{00}X_{00}^2 + g_{11}X_{11}^2 + \cdots + g_{dd}X_{dd}^2 = 0.$$

Since $F = \mathrm{GF}(2^e)$, every element of $F$ is a square as shown in Section 1.2. In particular, we may write $g_{ii} = a_{ii}^2$ for each $i$, for some element $a_{ii} \in F$. Hence, using the Frobenius automorphism $x \mapsto x^2$, we may write the above equation for the set of absolute points as

$$(a_{00}X_{00} + a_{11}X_{11} + \cdots + a_{dd}X_{dd})^2 = 0$$

and thus

$$a_{00}X_{00} + a_{11}X_{11} + \cdots a_{dd}X_{dd} = 0.$$

Since $g_{ii} \neq 0$ for some $i$, necessarily $a_{ii} \neq 0$ for that particular $i$, and we have a nontrivial linear equation describing the set of absolute points. That is, the absolute points form a hyperplane of $\mathrm{PG}(d, 2^e)$. □

Since we think of a hyperplane as being a "degenerate" example for a set of absolute points of an orthogonal polarity, when $\mathrm{char}(F) = 2$ the orthogonal polarities are typically called *pseudopolarities*, as previously mentioned.

We have yet to discuss the absolute points of the unitary polarities. In fact, these sets of absolute points are the main motivating force for this book.

**Definition 1.31.** *Let $\rho$ be a unitary polarity of the classical projective geometry* $\mathrm{PG}(d, F)$, *for some $d \geq 2$. The set of absolute points of $\rho$ is called a* **nondegenerate Hermitian variety***, and is denoted by* $\mathcal{H}(d, F)$. *If $d = 2$, the set of absolute points is called a* **nondegenerate Hermitian curve***.

Recall that for unitary polarities to exist, the field $F$ must admit an involutory automorphism. Hence, in the finite field case, we must have $F = \mathrm{GF}(q^2)$ for some prime power $q$, and we denote the Hermitian variety by $\mathcal{H}(d, q^2)$. In this case, as previously discussed, the involutory automorphism is $\sigma : x \mapsto x^q$.

It is important to note that over infinite fields admitting an involutory automorphism, Hermitian varieties may be empty. For instance, consider the field of complex numbers with complex conjugation as the involutory automorphism. If you take the identity matrix as the Gram matrix with respect to some fixed basis (it is a Hermitian matrix over the complex numbers), we obtain the equation

$$X_0\overline{X}_0 + X_1\overline{X}_1 + \cdots + X_d\overline{X}_d = 0$$

for the associated Hermitian variety, where $\overline{X}$ denotes the complex conjugate of $X$. But the only solution to this equation is the trivial solution, which yields no projective point. Hence this Hermitian variety is empty. Of course, one can obtain nonempty Hermitian varieties over the complex numbers by choosing different (Hermitian) Gram matrices.

Luckily, over finite fields all Hermitian varieties are nonempty, as was true for quadrics. Also, like quadrics, there are relatively few of them. In the next chapter we carefully study finite Hermitian curves.

We assume the following two facts from classical group theory. Let $\mathcal{H} = \mathcal{H}(d, q^2)$ be a finite nondegenerate Hermitian variety in $PG(d, q^2)$, for some $d \geq 2$. Let $PGU(d + 1, q^2)$ be the homography subgroup leaving $\mathcal{H}$ invariant.

1. The group $PGU(d + 1, q^2)$ acts transitively on the points of $\mathcal{H}$ and transitively on the points of $PG(d, q^2) \setminus \mathcal{H}$. That is, any point of $\mathcal{H}$ can be mapped to any other point of $\mathcal{H}$ by some homography that leaves $\mathcal{H}$ invariant as a set of points. A similar statement holds for points not on $\mathcal{H}$.
2. Any nondegenerate Hermitian variety of $PG(d, q^2)$ can be mapped to any other nondegenerate Hermitian variety of $PG(d, q^2)$ by some homography (projectivity) of $PG(d, q^2)$. We thus say that $\mathcal{H}$ is uniquely determined, up to **projective equivalence**.

**Theorem 1.32.** *Let $P = \langle v \rangle$ be some point of a nonempty, nondegenerate Hermitian variety $\mathcal{H} = \mathcal{H}(d, F)$, and let $Q = \langle w \rangle$ be some point of $P^\perp$, other than $P$, where $P^\perp$ denotes the orthogonal complement of the point $P$ with respect to the associated Hermitian form $s$. Then the line $PQ$ either meets $\mathcal{H}$ in precisely one point, necessarily $P$, or lies totally on $\mathcal{H}$.*

*Proof.* Since $P = \langle v \rangle$ is on $\mathcal{H}$, we know that $s(v, v) = 0$ and $P \in P^\perp$. Similarly, since $Q = \langle w \rangle \in P^\perp$, we know that $s(v, w) = 0$ and hence $s(w, v) = 0$. Let $R = \langle tv + w \rangle$ be an arbitrary point of $PQ \setminus \{P\}$. Then

$$s(tv + w, tv + w) = tt^\alpha s(v, v) + ts(v, w) + t^\alpha s(w, v) + s(w, w) = s(w, w).$$

Thus, if $s(w, w) = 0$, every point of the line $PQ$ lies on $\mathcal{H}$; otherwise, the line $PQ$ meets $\mathcal{H}$ only in the point $P$.     □

We conclude this opening chapter by discussing polarities in arbitrary finite projective planes, not necessarily classical. For a projective plane to admit a polarity, the plane must be **self-dual**; that is, the plane must be isomorphic to its dual plane. However, it is not known if every self-dual projective plane must admit a polarity. Suppose now that $\pi$ is some finite projective plane that admits a polarity $\rho$. Thus $\rho$ is a bijection of order two interchanging the points and lines of $\pi$ in such a way that incidence is preserved. That is, a point $P$ lies on a line $\ell$ of $\pi$ if and only if the point $\ell^\rho$ lies on the line $P^\rho$. Since $\pi$ is not necessarily classical, we cannot assume there is an underlying three-dimensional vector space over some field nor can we assume there is an associated sesquilinear form. However, using combinatorial arguments (and some linear algebra involving the incidence matrix of the plane), much still can be said in this general setting. The following results are due to Baer [17]. Elementary proofs may be found in Chapter 12 of [140].

**Theorem 1.33.** *Let $\rho$ be a polarity of a finite projective plane of order $n$. Then $\rho$ has at least $n + 1$ absolute points. If $\rho$ has exactly $n + 1$ absolute points, then*

1. *the absolute points are collinear and thus form a line when $n$ is even,*
2. *the absolute points form an oval when $n$ is odd.*

*Moreover, if $n$ is not a square, then $\rho$ must have exactly $n + 1$ absolute points.*

When the order of the finite plane is a square, an upper bound for the number of absolute points was proved by Seib [200]. Seib also showed that if this upper bound is achieved, the resulting set of absolute points is a very nice geometrical configuration. Purely combinatorial arguments for these results, which we state below, may be found in Chapter 12 of [140].

**Theorem 1.34.** *Let $\rho$ be a polarity of a finite projective plane of order $n = s^2$. Then the number of absolute points of $\rho$ is at most $s^3 + 1$. Moreover, if the number of absolute points is exactly $s^3 + 1$, then every line of the plane meets this set of absolute points in either 1 or $s + 1$ points.*

As we shall see in the next chapter, this upper bound is achievable in the classical (Desarguesian) projective plane $\mathrm{PG}(2, q^2)$ by choosing a unitary polarity $\rho$ arising from a Hermitian form on the underlying vector space. Thus we adopt the following terminology. In any square order projective plane containing a polarity $\rho$, we say that $\rho$ is a **unitary polarity** if the number of absolute points achieves the upper bound in Theorem 1.34. In Chapter 5 we will see examples of unitary polarities in non-Desarguesian projective planes.

# 2

# Hermitian Curves and Unitals

## 2.1 Nondegenerate Hermitian Curves

Let $\mathcal{H} = \mathcal{H}(2, q^2)$ be a nondegenerate Hermitian curve in the classical projective plane $\mathrm{PG}(2, q^2)$ of order $q^2$, for some prime power $q$. We begin this section by determining the combinatorial properties of this curve.

**Theorem 2.1.** *A nondegenerate Hermitian curve $\mathcal{H}$ in $\mathrm{PG}(2, q^2)$ has precisely $q^3 + 1$ points.*

*Proof.* As any two nondegenerate Hermitian curves are projectively equivalent, we may assume that $\mathcal{H}$ has equation

$$X_0^{q+1} + X_1^{q+1} + X_2^{q+1} = 0.$$

That is, we assume a basis for the underlying vector space has been chosen so that the Gram matrix for the associated Hermitian form is the identity matrix. Then the points of $\mathcal{H}$ have left-normalized homogeneous coordinates with precisely one of the following forms:

(i)  $(0, 1, z)$, $z^{q+1} = -1$,
(ii)  $(1, y, 0)$, $y^{q+1} = -1$,
(iii)  $(1, y, z)$, $y^{q+1} = f \neq -1$ and $z^{q+1} = -1 - f \neq 0$.

Using our discussion of norms from Section 1.2, we know that there are precisely $q + 1$ points of type (i) and $q + 1$ points of type (ii). Moreover, in the type (iii) case, there are precisely $q + 1$ values of $y$ in $\mathrm{GF}(q^2)$ with $y^{q+1} = -1$, and hence there are exactly $q^2 - q - 1$ possible second coordinates for any point of type (iii). Given any such second coordinate, there are exactly $q + 1$ choices for the third coordinate $z$, and hence there are precisely $(q + 1)(q^2 - q - 1)$ points of type (iii). Therefore the total number of points on $\mathcal{H}$ is $(q + 1)(q^2 - q + 1) = q^3 + 1$. $\square$

S. Barwick, G. Ebert, *Unitals in Projective Planes*,
DOI: 10.1007/978-0-387-76366-8_2, © Springer Science+Business Media, LLC 2008

Thus we see that a unitary polarity in a square order classical projective plane $PG(2, q^2)$, induced by a nondegenerate Hermitian form on the underlying vector space, does achieve the upper bound for the possible number of absolute points of a plane polarity given in Theorem 1.34. Thus we further know from Theorem 1.34 that the set of absolute points, namely, the Hermitian curve $\mathcal{H}$, has nice intersection properties with respect to all lines in $PG(2, q^2)$. However, we prefer to develop these intersection properties directly in this classical setting. Doing so will enable us to better understand Hermitian curves.

Let $\ell$ be any line of $PG(2, q^2)$. Then the pole $P = \ell^\perp$ of $\ell$ is either a point of $\mathcal{H} = \mathcal{H}(2, q^2)$ or it is not. Since $PGU(3, q^2)$ acts transitively on the points of $\mathcal{H}$ and also transitively on the points of $PG(2, q^2) \backslash \mathcal{H}$, we may assume not only our favorite form for $\mathcal{H}$, but also our favorite point on or off the curve $\mathcal{H}$. As above, take as our equation for $\mathcal{H}$ the *canonical equation*

$$X_0^{q+1} + X_1^{q+1} + X_2^{q+1} = 0.$$

Suppose first that $P = (0, 0, 1)$. Then $P \notin \mathcal{H}$ and thus $P$, as well as $P^\perp = \ell$, is nonabsolute. Using the identity matrix as our Gram matrix, we see that $\ell$ has dual coordinates $(0, 0, 1)'$ and hence equation $X_2 = 0$. Thus the homogeneous coordinates for any point of $\ell \cap \mathcal{H}$ are of the form $(1, y, 0)$, where $y^{q+1} = -1$. Hence, again using the properties of the norm function, we see that every nonabsolute line meets $\mathcal{H}$ in precisely $q + 1$ points. Often these lines are called **secant lines** of $\mathcal{H}$, and their intersections with $\mathcal{H}$ are called **chords**.

Next assume that $P = (0, 1, z)$, for some $z \in GF(q^2)$ with $z^{q+1} = -1$. Thus $P \in \mathcal{H}$ and hence $P$, as well as $\ell = P^\perp$, is absolute. This time the dual coordinates for $\ell$ are $(0, 1, z^q)'$, and thus $\ell$ has equation $X_1 + z^q X_2 = 0$. By Theorem 1.32, we know that either $\ell$ lies totally on $\mathcal{H}$ or $\ell$ meets $\mathcal{H}$ only in the point $P$. Since $(1, 0, 0)$ is a point of $\ell$ which is not on $\mathcal{H}$, $\ell$ meets $\mathcal{H}$ only in $P$. That is, every absolute line meets $\mathcal{H}$ in precisely one point, namely, its pole. We thus have the following result.

**Theorem 2.2.** *Let $\mathcal{H}$ be a nondegenerate Hermitian curve in $PG(2, q^2)$. Then every line of $PG(2, q^2)$ meets $\mathcal{H}$ in 1 or $q + 1$ points.*

In general, for a Hermitian variety $\mathcal{H} = \mathcal{H}(d, q^2)$ of dimension $d \geq 2$, a line meeting $\mathcal{H}$ in one point will be called a **tangent line** of the variety, and a line meeting $\mathcal{H}$ in $q + 1$ points will be called a **hyperbolic line** (or **secant line**) of the variety. Thus, for nondegenerate Hermitian curves $\mathcal{H}$ embedded in $PG(2, q^2)$, every line of $PG(2, q^2)$ is a tangent line or a hyperbolic line of $\mathcal{H}$. In particular, no line of $PG(2, q^2)$ is disjoint from $\mathcal{H}$ and no line of $PG(2, q^2)$ lies totally on the curve $\mathcal{H}$. As indicated above, the hyperbolic lines of a nondegenerate Hermitian curve are often called secant lines of the curve.

**Theorem 2.3.** *Let $\mathcal{H}$ be a nondegenerate Hermitian curve in $\mathrm{PG}(2,q^2)$. Then $\mathcal{H}$ has $q^3+1$ tangent lines and $q^4-q^3+q^2$ secant lines in $\mathrm{PG}(2,q^2)$. In particular, through each point of $\mathcal{H}$ there pass $q^2$ secant lines and one tangent line, while through each point of $\mathrm{PG}(2,q^2)\backslash\mathcal{H}$ there pass $q+1$ tangent lines and $q^2-q$ secant lines (see Figure 2.1).*

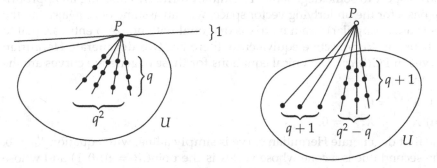

**Fig. 2.1.** The number of tangents and secants through a point $P$ on and off a unital

*Proof.* The line through any two distinct points of $\mathcal{H}$ must be a secant line by Theorem 2.2. Furthermore, $\mathcal{H}$ has $q^3+1$ points by Theorem 2.1. Hence the number of distinct secant lines of $\mathcal{H}$ is

$$\binom{q^3+1}{2}\Big/\binom{q+1}{2} = q^2(q^2-q+1) = q^4-q^3+q^2$$

(since every secant line is counted $\binom{q+1}{2}$ times). Since $\mathrm{PG}(2,q^2)$ has $q^4+q^2+1$ lines, all of which are tangent or secant, we see that $\mathcal{H}$ has $q^3+1$ tangent lines. In particular, through each point $P \in \mathcal{H}$ there is precisely one tangent line, namely, $P^\perp$, and hence $q^2$ secant lines.

If $Q$ is any point of $\mathrm{PG}(2,q^2)\backslash\mathcal{H}$, let $t$ be the number of tangent lines through $Q$ and let $h$ be the number of secant lines through $Q$. Then

$$t+h = q^2+1 \quad \text{and} \quad t\cdot 1+h\cdot(q+1) = q^3+1,$$

implying that $h = q^2-q$ and $t = q+1$.     □

For any point $Q \notin \mathcal{H}$, the $q+1$ points of $\mathcal{H}$ lying on the $q+1$ tangent lines through $Q$ are often called the **feet** of $Q$.

**Corollary 2.4.** *Let $\mathcal{H}$ be a nondegenerate Hermitian curve in $\mathrm{PG}(2,q^2)$. If $Q$ is any point of $\mathrm{PG}(2,q^2)\backslash\mathcal{H}$, then the feet of $Q$ are collinear, lying on the line $Q^\perp$.*

*Proof.* Let $\ell$ be a tangent line of $\mathcal{H}$ through $Q$, and let $P = \ell\cap\mathcal{H}$. Then $P^\perp = \ell$ and hence $\ell^\perp = P$. But $Q \in \ell$ implies that $P = \ell^\perp \in Q^\perp$. That is, the $q+1$ feet of $Q$ all lie on $Q^\perp$. In fact, as $Q^\perp$ is a secant line, the feet of $Q$ consist precisely of the points in $Q^\perp\cap\mathcal{H}$.     □

## 2.2 Degenerate Hermitian Curves and Baer Sublines

While the absolute points of a unitary polarity of $PG(2, q^2)$ form a nondegenerate Hermitian curve by definition, one can also discuss "degenerate" Hermitian curves. Namely, we look at the projective points determined by the nonzero vectors in the underlying vector space that are self-orthogonal with respect to some degenerate Hermitian form. By choosing an appropriate basis for the underlying vector space, we can assume once again that the associated (singular) Gram matrix is diagonal with nonzero entries equal to 1. Thus, up to projective equivalence, there are two **degenerate Hermitian curves** in $PG(2, q^2)$. Canonical equations for these degenerate curves are the following:

(i) $X_0^{q+1} = 0$,

(ii) $X_0^{q+1} + X_1^{q+1} = 0$.

The first degenerate Hermitian curve is simply a line, with equation $X_0 = 0$. The second one is a cone whose vertex is the point $P = (0, 0, 1)$ and whose base can be taken to be the set of points $B = \{(1, y, 0) \mid y^{q+1} = -1\}$ on the line with equation $X_2 = 0$. In fact, one may think of $B$ as a nondegenerate Hermitian variety of the projective line $PG(1, q^2)$, whose underlying vector space is a two-dimensional vector space over $GF(q^2)$. This makes perfectly good sense, even though in the previous chapter we only defined Hermitian varieties for projective dimension at least two.

**Theorem 2.5.** *Let $\mathcal{H}(1, q^2)$ be a nondegenerate Hermitian variety of the projective line $PG(1, q^2)$. Then $\mathcal{H}(1, q^2)$ is isomorphic to $PG(1, q)$, the projective line over the subfield $GF(q)$ of $GF(q^2)$.*

*Proof.* Just as in projective dimensions at least two, there is a unique nondegenerate Hermitian variety in $PG(1, q^2)$, up to projective equivalence. Instead of taking the canonical equation (ii) above, we take an embedding of $\mathcal{H}(1, q^2)$ in $PG(1, q^2)$ that is in nonstandard position. If $q$ is odd and $\zeta$ is a primitive element of $GF(q^2)$, we choose $\epsilon = \zeta^{(q+1)/2}$ and let our Gram matrix be

$$H = \begin{pmatrix} 0 & \epsilon \\ \epsilon^q & 0 \end{pmatrix}.$$

Note that $H$ is Hermitian since $\epsilon^{q-1} = \zeta^{(q^2-1)/2} = -1$ and hence $\epsilon^q = -\epsilon$. Thus the equation for $\mathcal{H}(1, q^2)$ becomes

$$\epsilon X_0 X_1^q - \epsilon X_0^q X_1 = 0 \quad \text{or} \quad X_0 X_1^q - X_0^q X_1 = 0.$$

For even $q$, we may take

$$H = \begin{pmatrix} 0 & 1 \\ 1 & 0 \end{pmatrix}$$

as our nonsingular Hermitian Gram matrix, thereby obtaining the equation $X_0 X_1^q + X_0^q X_1 = 0$ for $\mathcal{H}(1, q^2)$. Since $1 = -1$ when $q$ is even, we obtain the same equation for $\mathcal{H}(1, q^2)$ in both cases. Moreover, using this equation, we see that

$$\begin{aligned} \mathcal{H}(1, q^2) &= \{(0, 1)\} \cup \{(1, y) \mid y^q = y\} \\ &= \{(0, 1)\} \cup \{(1, y) \mid y \in \mathrm{GF}(q)\} \\ &\cong \mathrm{PG}(1, q), \end{aligned}$$

and the proof is complete.    □

At this stage one should more carefully discuss our model for the (classical) projective line. Using left-normalized coordinates, as introduced in the previous chapter, one may identify the points of $L = \mathrm{PG}(1, q^2)$ with $\{(0, 1)\} \cup \{(1, y) \mid y \in \mathrm{GF}(q^2)\}$. Using the subfield $\mathrm{GF}(q)$ of $\mathrm{GF}(q^2)$, the set of points

$$B = \{(0, 1)\} \cup \{(1, y) \mid y \in \mathrm{GF}(q)\}$$

is a copy of the subline $\mathrm{PG}(1, q)$ in **standard position** on the line $L$. Any element in the orbit of $B$ under the homography subgroup $\mathrm{PGL}(2, q^2)$ of $\mathrm{PG}(1, q^2)$ is defined to be a **Baer subline** of $L = \mathrm{PG}(1, q^2)$.

In particular, the above theorem says that $\mathcal{H}(1, q^2)$ is a Baer subline of $\mathrm{PG}(1, q^2)$. Similarly, any secant line of the nondegenerate Hermitian curve $\mathcal{H}(2, q^2)$ meets $\mathcal{H}(2, q^2)$ in a Baer subline, just as the base for the degenerate Hermitian curve of type (ii) is a Baer subline.

One particularly useful fact, which we prove below, is that a Baer subline is uniquely determined by any three of its points.

**Theorem 2.6.** *Let $P$, $Q$, and $R$ be three distinct points of $\mathrm{PG}(1, q^2)$. Then there is a unique Baer subline of $\mathrm{PG}(1, q^2)$ containing $P$, $Q$, and $R$.*

*Proof.* Write $P = \langle v \rangle$ and $Q = \langle w \rangle$ for some nonzero vectors of the underlying two-dimensional vector space $V$ for $\mathrm{PG}(1, q^2)$. Since $P$ and $Q$ are distinct points, $v$ and $w$ are linearly independent vectors and hence form a basis for $V$. Since the point $R$ is distinct from both $P$ and $Q$, we may write $R = \langle v + \lambda w \rangle$ for some nonzero element $\lambda$ in $\mathrm{GF}(q^2)$. Replacing $w$ by $u = \lambda w$, we then have $Q = \langle u \rangle$ and $R = \langle v + u \rangle$. Thus

$$\{\langle u \rangle\} \cup \{\langle v + tu \rangle \mid t \in \mathrm{GF}(q)\}$$

is a Baer subline of $\mathrm{PG}(1, q^2)$ containing $P$, $Q$, and $R$. The uniqueness follows from the uniqueness of $\mathrm{GF}(q)$ as a subfield of order $q$ in $\mathrm{GF}(q^2)$.    □

The notion of Baer subline may be generalized to higher dimensions. For the purposes of this book, we only need to do this for projective dimension two.

**Definition 2.7.** *A* **Baer subplane** $\mathcal{B}$ *of* $\mathrm{PG}(2, q^2)$ *is a subset of points and (partial) lines of* $\mathrm{PG}(2, q^2)$ *forming an isomorphic copy of* $\mathrm{PG}(2, q)$. *In general, if* $\mathcal{P}$ *is any projective plane of order* $n^2$, *a* **Baer subplane** $\mathcal{B}$ *of* $\mathcal{P}$ *is a subset of points and (partial) lines of* $\mathcal{P}$ *forming a projective plane of order* $n$. *The projective plane* $\mathcal{B}$ *may or may not be classical. A* **Baer subline** $b$ *of* $\mathcal{P}$ *is the intersection of a Baer subplane* $\mathcal{B}$ *of* $\mathcal{P}$ *with a line of* $\mathcal{P}$ *that meets* $\mathcal{B}$ *in* $n + 1$ *points (that is, $b$ is a line of* $\mathcal{B}$).

It should be noted that if $\mathcal{B}$ is a Baer subplane of the projective plane $\mathcal{P}$ of order $n^2$, then any line $\ell$ of $\mathcal{P}$ which contains two points of $\mathcal{B}$ must necessarily intersect $\mathcal{B}$ in the unique line of $\mathcal{B}$ determined by those two points. That is, such a line $\ell$ must contain precisely $n + 1$ points of $\mathcal{B}$. The following result in the classical projective plane $\mathrm{PG}(2, q^2)$ is analogous to Theorem 2.6.

**Theorem 2.8.** *Let* $P, Q, R$, *and* $S$ *be four distinct points of* $\mathrm{PG}(2, q^2)$, *with no three collinear. Then there is a unique Baer subplane of* $\mathrm{PG}(2, q^2)$ *containing* $P, Q, R$, *and* $S$.

*Proof.* Using the underlying three-dimensional vector space $V$ for $\mathrm{PG}(2, q^2)$, basis vectors for the noncollinear points $P, Q$, and $R$ will be linearly independent and hence together will form a basis for $V$. Multiplying each of these basis vectors by an appropriate scalar, as in the proof of Theorem 2.6, we may write

$$P = \langle u \rangle, \quad Q = \langle v \rangle, \quad R = \langle w \rangle, \quad S = \langle u + v + w \rangle.$$

Thus $\{\langle w \rangle\} \cup \{\langle v + tw \rangle \mid t \in \mathrm{GF}(q)\} \cup \{\langle u + sv + tw \rangle \mid s, t \in \mathrm{GF}(q)\}$ is a Baer subplane containing $P, Q, R$, and $S$. The uniqueness follows as in Theorem 2.6. $\qquad\square$

Using the idea of the above proof, given any quadrangle $P, Q, R, S$ of points in $\mathrm{PG}(2, q^2)$, we may choose coordinates so that a basis vector for $S$ is the sum of the basis vectors for $P, Q$, and $R$, respectively. Hence, given any two quadrangles of $\mathrm{PG}(2, q^2)$, one can find an element of $\mathrm{PGL}(3, q^2)$ (some homography) mapping one quadrangle to the other. That is, using Theorem 2.8, we see that $\mathrm{PGL}(3, q^2)$ acts transitively on Baer subplanes of $\mathrm{PG}(2, q^2)$, just as $\mathrm{PGL}(2, q^2)$ acts transitively on Baer sublines of $\mathrm{PG}(1, q^2)$.

It is important to note that Theorem 2.8 does not hold in arbitrary square order projective planes. Various implications of this fact will be apparent in the next chapter when we study the Bruck-Bose representation for a certain family of square order planes which are not necessarily classical.

Returning to the classical setting, we next consider the possible intersections of a nondegenerate Hermitian curve and a Baer subplane embedded in $\mathrm{PG}(2, q^2)$.

**Theorem 2.9.** *A nondegenerate Hermitian curve* $\mathcal{H}$ *and a Baer subplane* $\mathcal{B}$ *in* $\mathrm{PG}(2, q^2)$ *meet in* $1, q + 1$, *or* $2q + 1$ *points. More precisely, this intersection is a point, a line, two intersecting lines, or an irreducible conic in* $\mathcal{B}$.

*Proof.* Since $\text{PGL}(3, q^2)$ acts transitively on Baer subplanes of $\text{PG}(2, q^2)$, we may assume without loss of generality that the Baer subplane in question is in standard position, so that

$$\mathcal{B} = \{(0,0,1)\} \cup \{(0,1,z) \mid z \in \text{GF}(q)\} \cup \{(1,y,z) \mid y, z \in \text{GF}(q)\}.$$

Consider a general equation for $\mathcal{H}$, say

$$aX_0^{q+1} + bX_1^{q+1} + cX_2^{q+1} + \alpha X_0 X_1^q + \beta X_0 X_2^q$$
$$+ \gamma X_1 X_2^q + \alpha^q X_0^q X_1 + \beta^q X_0^q X_2 + \gamma^q X_1^q X_2 = 0,$$

where $a, b, c \in \text{GF}(q)$ and $\alpha, \beta, \gamma \in \text{GF}(q^2)$. Then $\mathcal{H} \cap \mathcal{B}$ consists of those points whose homogeneous coordinates satisfy

$$aX_0^2 + bX_1^2 + cX_2^2 + (\alpha + \alpha^q)X_0 X_1 + (\beta + \beta^q)X_0 X_2 + (\gamma + \gamma^q)X_1 X_2 = 0.$$

Since this is a quadratic equation over the subfield $\text{GF}(q)$, we see that $\mathcal{H} \cap \mathcal{B}$ is a conic (possibly degenerate) in $\mathcal{B} \cong \text{PG}(2, q)$. The result now follows from our discussion of conics in Section 1.3. $\qquad\qquad\square$

It should be noted that if both $\mathcal{H}$ and $\mathcal{B}$ are in standard (or canonical) position, then $\mathcal{H} \cap \mathcal{B}$ consists of all points satisfying the equation

$$X_0^2 + X_1^2 + X_2^2 = 0.$$

This intersection is an irreducible conic in $\text{PG}(2, q)$ if $q$ is odd, while it is a line if $q$ is even (since then $x \mapsto x^2$ is a field automorphism). This is indicative of a more general result. Namely, the automorphic collineation of $\text{PG}(2, q^2)$ induced by the conjugation mapping $\sigma : x \mapsto x^q$ has as its fixed points the Baer subplane $\mathcal{B}$ in standard position. This mapping of order two is called the associated **Baer involution** of $\mathcal{B}$. Since $\text{PGL}(3, q^2)$ acts transitively on Baer subplanes, we see that every Baer subplane of $\text{PG}(2, q^2)$ has an associated (nonlinear) Baer involution, whose fixed points are precisely the given Baer subplane. We state the following result of Seib [200] without proof.

**Theorem 2.10.** *Let $\mathcal{H}$ be a nondegenerate Hermitian curve in $\text{PG}(2, q^2)$ with associated unitary polarity $\rho$, and let $\mathcal{B}$ be a Baer subplane of $\text{PG}(2, q^2)$ with associated Baer involution $\phi$. If $\rho$ and $\phi$ commute as maps, then $\mathcal{H} \cap \mathcal{B}$ is a conic of $\mathcal{B}$ when $q$ is odd and a line of $\mathcal{B}$ when $q$ is even.*

## 2.3 Unitals

We begin this section with the notion of a *design*, a concept that first arose in the field of statistics but is now an integral part of discrete mathematics.

**Definition 2.11.** *Let $t, v, k,$ and $\lambda$ be positive integers with $t < k < v$. A **$t$-$(v, k, \lambda)$** **design** is an ordered pair $(\mathcal{V}, \mathcal{B})$, where $\mathcal{V}$ is set of $v$ elements called **points** and $\mathcal{B}$ is a collection of certain subsets of $\mathcal{V}$ of size $k$ called **blocks**, such that every subset of $t$ points of $\mathcal{V}$ is contained in exactly $\lambda$ blocks.*

**Definition 2.12.** *Two designs are called **isomorphic** if there is a bijection from the points and blocks of one design to the points and blocks of the other design that preserves containment.*

Note that a finite projective plane of order $n$ is a 2-$(n^2 + n + 1, n + 1, 1)$ design, where the blocks are the lines of the projective plane. In fact, a simple counting argument shows that any design with these parameters is necessarily a projective plane of order $n$, whose lines are the blocks of the design. Similarly, an affine plane of order $n$ is a 2-$(n^2, n, 1)$ design, and any design with these parameters is necessarily an affine plane of order $n$.

Now let $\mathcal{H} = \mathcal{H}(2, q^2)$ be a nondegenerate Hermitian curve in $\mathrm{PG}(2, q^2)$. Any two distinct points of $\mathcal{H}$ uniquely determine a line of $\mathrm{PG}(2, q^2)$, which is necessarily a secant line meeting $\mathcal{H}$ in $q + 1$ points. Hence, if we take the $q^3 + 1$ points of $\mathcal{H}$ as the points of our design and take all the secant line intersections with $\mathcal{H}$ (that is, the chords of $\mathcal{H}$) as our blocks, we obtain a 2-$(q^3 + 1, q + 1, 1)$ design. Ignoring the projective plane $\mathrm{PG}(2, q^2)$ and concentrating solely on the above design, we make the following definition.

**Definition 2.13.** *Let $n$ be an integer, $n \geq 3$. A **unital** of **order** $n$ is any 2-$(n^3 + 1, n + 1, 1)$ design.*

Note that if $n = 2$, then a 2-$(9, 3, 1)$ design is an affine plane of order 3. Thus we require $n \geq 3$ in our definition of unital. However, we do not require that $n$ be a prime power, nor do we require that a unital of order $n$ be embedded in a projective plane of order $n^2$. If such an embedding exists, so that the blocks of the unital $U$ are collinear sets of points and thus every line of the ambient plane meets $U$ in 1 or $n + 1$ points, then we call $U$ an **embedded unital**. Of course, the plane of order $n^2$ in which $U$ is embedded may or may not be classical. In any case, the combinatorial results of Theorem 2.3 remain true for any embedded unital, with $q$ replaced by $n$. However, Corollary 2.4 does not hold for unitals in general, since the unital may not arise as the absolute points of some unitary polarity on the ambient square order plane, and hence the feet of some points off an embedded unital might not be collinear.

Nonetheless, if a unital $U$ of order $n$ is embedded in some projective plane $\mathcal{P}$ of order $n^2$, then any two of the $n^3 + 1$ tangent lines to $U$ in $\mathcal{P}$ necessarily meet in a point of $\mathcal{P} \setminus U$. Hence we can define a new design $U^*$ by taking the tangent lines to $U$ as the *points* of $U^*$ and the points of $\mathcal{P} \setminus U$ as the *blocks* of $U^*$; incidence is given by reverse containment. Since each point of $\mathcal{P} \setminus U$ lies on $n + 1$ tangent lines to $U$, we obtain another 2-$(n^3 + 1, n + 1, 1)$ design. That is, $U^*$ is another unital, called the **dual unital** to $U$. It is embeddable in the dual plane $\mathcal{P}^*$. If $U$ and $U^*$ happen to be isomorphic as designs, then $U$ is called **self-dual**. Of course, one can only speak of the dual unital $U^*$ if the unital $U$ of order $n$ is embedded in some projective plane of order $n^2$.

The nondegenerate Hermitian curve $\mathcal{H}(2, q^2)$ is often called the **classical unital** of order $q$. There are also many nonclassical unitals of order $q$, for any prime power $q$, as we shall soon see. So far only one unital has been constructed whose order is not a prime power; namely, a unital of order 6 constructed by Mathon [169] and independently by Bagchi and Bagchi [19]. It is still unknown if this unital can be embedded in a projective plane of order 36 (see [142, 143]), assuming there is such a plane.

Another very interesting family of unitals are the **Ree unitals** of order $q$. These unitals admit a collineation group isomorphic to the classical Ree group of order $(q^3 + 1)q^3(q - 1)$, where $q$ is necessarily an odd power of 3. It was shown by Lüneburg [166] that the Ree unital cannot be embedded in any projective plane in such a way that the group carries over. For $q = 3$, Grüning [126] gave a geometric proof that the Ree unital of order 3 cannot be embedded in a projective plane of order 9.

We next discuss the notion of *equivalence* for unitals embedded in the same projective plane. If treated solely as designs, ignoring the ambient plane in which the unitals are embedded, then we say two such unitals are **isomorphic** provided they are isomorphic as designs, using Definition 2.12. However, we say two such unitals are **equivalent** if there is a collineation of the ambient plane that maps one unital to the other. This is consistent with our previous convention for classical unitals (Hermitian curves). Namely, we said any two Hermitian curves are *projectively equivalent* because there is always a projectivity (a certain type of collineation acting on the ambient classical plane) that maps one Hermitian curve to another.

As an illustration of this idea, recall that there are precisely four projective planes of order 9, one of which is classical. The other three are the Hall plane of order 9, the dual Hall plane of order 9, and the Hughes plane of order 9 (which is a self-dual projective plane). We will meet all these projective planes later in the book. An exhaustive computer search in [178] found that there are 18 unitals (up to equivalence) of order 3 embedded in these four planes. Namely, there are two mutually inequivalent unitals in the classical (Desarguesian) plane of order 9, four mutually inequivalent unitals in the Hall plane of order 9 (hence four mutually inequivalent unitals in the dual Hall plane of order 9), and eight mutually inequivalent unitals in the Hughes plane of order 9. However, one of the unitals in the Hall plane is self-dual as a design, and hence is counted in both the Hall plane and the dual Hall plane. Thus there are at most 17 mutually nonisomorphic unitals, as designs, in this list.

On the other hand, a nonexhaustive computer search in [43] found over 900 mutually nonisomorphic designs which are unitals of order 3. Hence it appears that the vast majority of unitals are not embedded in square order projective planes.

Nonetheless, we will restrict our attention to those unitals of order $n$ which are embedded in a (possibly nonclassical) projective plane of order $n^2$, since the geometry is most interesting in this case.

To further motivate the study of embedded unitals, we define the notion of a *blocking set*. Most likely, the first appearance of blocking sets in the mathematical literature occurred in connection with game theory; in particular, with projective games (see [186], [220]). In such a setting, a blocking set is a collection of points that meets every winning coalition but contains no minimal winning coalition. In more recent times, blocking sets generally are defined in finite projective and affine spaces of arbitrary dimension, where they have numerous connections and applications to various topics in finite geometry. Here we only define blocking sets in (finite) projective planes.

**Definition 2.14.** *A* **blocking set** *in a projective plane* $\mathcal{P}$ *is a subset of points in* $\mathcal{P}$ *that meets every line of* $\mathcal{P}$ *but contains no line of* $\mathcal{P}$. *A blocking set is called* **reduced** *(or* **minimal***) if no proper subset of it is also a blocking set.*

Note that we require the blocking set not to contain a line in order to avoid trivial examples. The following fundamental result was proved by Bruen. The original proof separated the square and nonsquare cases for the order of the plane, and may be found in [75], [76]. However, a much simpler proof covering all cases may be found in [79].

**Theorem 2.15.** *Any blocking set in a projective plane* $\mathcal{P}$ *of order* $n$ *has at least* $n + \sqrt{n} + 1$ *points. Moreover, if this lower bound is met, then* $n$ *is necessarily square and the blocking set consists of the points in some Baer subplane of* $\mathcal{P}$.

For the purposes of this book, perhaps the following upper bound is more significant. This upper bound for the size of a minimal blocking set was proved by Bruen and Thas (see [80] for the proof).

**Theorem 2.16.** *Any minimal blocking set in a projective plane* $\mathcal{P}$ *of order* $n$ *has at most* $n\sqrt{n} + 1$ *points. Moreover, if this upper bound is reached, then* $n$ *is necessarily square and the blocking set consists of the points of some unital embedded in* $\mathcal{P}$.

Thus we see that unitals play a parallel role to Baer subplanes when considering extreme values for the size of a blocking set in a square order projective plane. The mere fact that unitals are blocking sets adds further credence to our belief that unitals play a key role in understanding the nature of all square order projective planes.

In the following two chapters we develop a general method for constructing unitals embedded in a broad family of projective planes of order $q^2$, where $q$ is a prime power. In general, these unitals do not consist of the absolute points of some unitary polarity defined on the ambient projective plane. In Chapter 5 we discuss different techniques for constructing unitals in a wide variety of non-Desarguesian square order projective planes. In particular, we construct unitals of order $q$ which are embeddable in two nonisomorphic projective planes of order $q^2$. In Chapter 6 we make an in-depth investigation of the combinatorial structure of embedded unitals and discuss certain associated combinatorial objects. Finally, in Chapter 7 we present numerous combinatorial and geometric characterizations of unitals embedded in the classical plane $PG(2, q^2)$. We conclude with a list of open problems in Chapter 8.

# 3

# Translation Planes

In this chapter, we look at a special class of projective planes called translation planes. We restrict our attention to the finite plane case, although the concepts introduced here generalize to the infinite case. We show that any translation plane of order $q^2$ with "kernel" containing $\mathrm{GF}(q)$ can be represented in $\mathrm{PG}(4, q)$. This representation is very useful for studying unitals in these planes. In particular, it can be used to study unitals embedded in the Desarguesian plane.

## 3.1 Translation Planes

Recall that a collineation $\phi$ of a projective plane $\mathcal{P}$ is a bijection from points to points and lines to lines that preserves incidence.

**Definition 3.1.** *A nonidentity collineation $\phi$ is called a* **central collineation** *(or* **perspectivity***) if*

1. *$\phi$ fixes a point $P$ linewise (that is, $m^\phi = m$ for all lines $m$ incident with $P$),*
2. *$\phi$ fixes a line $\ell$ pointwise (that is, $Q^\phi = Q$ for all points $Q \in \ell$).*

More precisely, such a map $\phi$ is called a $(P, \ell)$-central collineation or a $(P, \ell)$-perspectivity. The point $P$ is uniquely determined and is called the **center** of $\phi$, while the uniquely determined line $\ell$ is called the **axis** of $\phi$. If $P \in \ell$, then $\phi$ is called an **elation**; otherwise, $\phi$ is called a **homology**. The center and the points on the axis are the only points fixed by $\phi$. Thus, if $\phi$ is an elation, then in the affine plane $\mathcal{P} \backslash \ell$, the mapping $\phi$ is a **translation** (that is, a fixed-point-free collineation preserving parallel classes). Thus sometimes elations are also called translations, and it is these collineations in which we are most interested.

The plane $\mathcal{P}$ is called $(P, \ell)$-**transitive** if for all points $X, Y$ of $\mathcal{P}$ with $X, Y, P$ a collinear triple of distinct points and $X, Y \notin \ell$, there exists a $(P, \ell)$-central collineation $\phi$ such that $X^\phi = Y$. If $\mathcal{P}$ is $(P, \ell)$-transitive for all points

S. Barwick, G. Ebert, *Unitals in Projective Planes*,
DOI: 10.1007/978-0-387-76366-8_3, © Springer Science+Business Media, LLC 2008

$P \in \ell$, then $\ell$ is a **translation line** of $\mathcal{P}$ and $\mathcal{P}$ is a **translation plane** with respect to $\ell$. If no line is specifically mentioned and $\mathcal{P}$ arose naturally from an affine plane, then we assume $\mathcal{P}$ is a translation plane with respect to $\ell_\infty$, the line at infinity. A well-known example of a translation plane is the Desarguesian plane $PG(2, q)$, which is a translation plane with respect to every line. It can be shown that if a finite translation plane has two distinct translation lines, then it is a Desarguesian plane (for instance, see Theorem 6.18 in [140]). Hence a non-Desarguesian translation plane has a unique translation line.

Every translation plane can be coordinatized by an algebraic structure called a **quasifield** (see Chapter 6 of [140]). Moreover, every quasifield has an algebraic substructure called its **kernel**, which is a skewfield over which the quasifield may be regarded as a (right or left) vector space. In the finite case this skewfield is necessarily a field. In particular, a finite translation plane $\mathcal{P}$ of order $q^2$ is coordinatized by some quasifield $Q$ of order $q^2$. If the kernel of $Q$ is $GF(q^2)$, then $Q = GF(q^2)$, $Q$ is a vector space of dimension one over its kernel, and $\mathcal{P} = PG(2, q^2)$. If $Q$ has kernel $GF(q)$, then $Q$ is a vector space of dimension two over its kernel and $\mathcal{P}$ is a non-Desarguesian translation plane. In this case we say $\mathcal{P}$ *has order $q^2$ and kernel* $GF(q)$, or $\mathcal{P}$ *has dimension two over its kernel*. In this book we will encounter several such examples, including the **Hall plane** of order $q^2$, denoted $\mathrm{Hall}(q^2)$. It is a translation plane coordinatized by the **Hall quasifield** of order $q^2$, which has kernel $GF(q)$. For more details on translation planes, including the Hall plane, see [140].

We also mention that elations and homologies generalize to higher dimensions. For instance, in $PG(3, q)$ an elation is a collineation fixing some plane $\pi$ pointwise and fixing all the lines through some point $P \in \pi$. If $(P, \pi)$ is any incident point-plane pair in $PG(3, q)$ and if $A, B$ are any two points of $PG(3, q)$, neither on $\pi$ and neither equal to $P$, with $P, A, B$ collinear, then there is always an elation of $PG(3, q)$ with axis $\pi$ and center $P$ that maps $A$ to $B$. That is, since $PG(3, q)$ is a Desarguesian space, there are as many elations as possible.

## 3.2 Derivation

In this section we describe a method for constructing new translation planes from *derivable* translation planes of order $q^2$. These planes contain certain Baer subplanes which are fundamental to the construction. We begin with some elementary counting concerning Baer subplanes.

**Theorem 3.2.** *Let $\mathcal{P}$ be a projective plane of order $q^2$, and let $\mathcal{B}$ be a Baer subplane of $\mathcal{P}$. A point $Q \notin \mathcal{B}$ lies on one line that meets $\mathcal{B}$ in a Baer subline and $q^2$ lines that meet $\mathcal{B}$ in a point. In particular, there are no lines in $\mathcal{P}$ which are disjoint from $\mathcal{B}$.*

*Proof.* Let $Q$ be a point of $\mathcal{P} \setminus \mathcal{B}$. For each point $P \in \mathcal{B}$, $PQ$ meets $\mathcal{B}$ in at least one point. As the number of lines through $Q$ is less than the number of points in $B$, necessarily $Q$ lies on at least one Baer subline of $\mathcal{B}$. Note that any line meeting $\mathcal{B}$ in at least two points must necessarily meet $\mathcal{B}$ in $q + 1$ points. If $Q$ were on two Baer sublines of $\mathcal{B}$, then since these Baer sublines meet in $Q$, we would have $Q \in \mathcal{B}$, a contradiction. Thus $Q$ lies on exactly one Baer subline of $\mathcal{B}$. As the $q^2 + 1$ lines through $Q$ partition the $q^2 + q + 1$ points of $\mathcal{B}$, we see that $Q$ lies on one Baer subline, $q^2$ tangents to $\mathcal{B}$, and no lines external to $\mathcal{B}$. $\qquad\qquad\square$

Let $\mathcal{A}$ be a finite affine plane of order $q^2$. Recall that we can uniquely complete $\mathcal{A}$ to a projective plane $\mathcal{P}$ by adding a slope point (or point at infinity) to each affine line, and then adding a line $\ell_\infty$ which contains all the slope points. If $\mathcal{B}$ is a Baer subplane of $\mathcal{P}$ which meets $\ell_\infty$ in a Baer subline, then $\mathcal{B} \setminus \ell_\infty$ is called an **affine Baer subplane** of $\mathcal{A}$. Now suppose that there exists a set $D$ of $q + 1$ points on $\ell_\infty$ such that for any two distinct points $X, Y$ of $\mathcal{A}$ for which $XY$ meets $\ell_\infty$ in a point of $D$, there is a unique Baer subplane of $\mathcal{P}$ containing $X, Y$, and $D$ (see Figure 3.1). In such a situation we call $D$

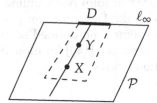

**Fig. 3.1.** A derivation set $D$

a **derivation set** of $\mathcal{A}$, and we say that $\mathcal{A}$ is a **derivable plane**. We use this derivation set to define a new incidence structure $\mathcal{D}(\mathcal{A})$ as follows:

- the *points* of $\mathcal{D}(\mathcal{A})$ are the points of $\mathcal{A}$;
- the *lines* of $\mathcal{D}(\mathcal{A})$ are
  (i) the lines of $\mathcal{A}$ whose projective extensions meet $\ell_\infty$ in a point not in $D$,
  (ii) the affine Baer subplanes of $\mathcal{A}$ whose projective extensions contain $D$;
- *incidence* is inclusion.

**Theorem 3.3.** *The above incidence structure $\mathcal{D}(\mathcal{A})$ is an affine plane of order $q^2$, called the* **derived plane** *of $\mathcal{A}$.*

*Proof.* The set $\mathcal{D}(\mathcal{A})$ contains $q^4$ points, and every line of $\mathcal{D}(\mathcal{A})$ contains $q^2$ points. Now let $P, Q$ be two points of $\mathcal{D}(\mathcal{A})$. Then in $\mathcal{P}$, the line $\ell = PQ$ meets $\ell_\infty$ in a point. If $PQ \cap \ell_\infty \notin D$, then $\ell$ corresponds to a line of $\mathcal{D}(\mathcal{A})$, and the point set $\{P, Q\}$ is not contained in a Baer subplane through $D$. Hence the points $P$ and $Q$ are contained in a unique line of $\mathcal{D}(\mathcal{A})$, which is of type (i). If $PQ \cap \ell_\infty \in D$, then since $D$ is a derivation set, there is a unique Baer subplane containing $P, Q$, and $D$. Hence in this case the points $P$ and $Q$ are contained in a unique line of $\mathcal{D}(\mathcal{A})$, which is of type (ii). Thus any two points of $\mathcal{D}(\mathcal{A})$ are contained in a unique line of $\mathcal{D}(\mathcal{A})$. Hence $\mathcal{D}(\mathcal{A})$ is a 2-$(q^4, q^2, 1)$ design. An elementary counting argument, as discussed in Section 2.3, now shows that $\mathcal{D}(\mathcal{A})$ is an affine plane of order $q^2$. □

We can uniquely complete $\mathcal{D}(\mathcal{A})$ to a projective plane $\mathcal{D}(\mathcal{P})$ by adding a line at infinity $\ell'_\infty$. The points of $\ell_\infty \setminus D$ have a natural correspondence to $q^2 - q$ points of $\ell'_\infty$. We denote the remaining $q + 1$ points of $\ell'_\infty$ by $D'$, and note that they do not correspond in any natural way to the points of $D$. Also note that $D'$ is a derivation set of $\mathcal{D}(\mathcal{A})$, and deriving with respect to $D'$ yields the original affine plane $\mathcal{A}$. Derivation is an affine process, but since every affine plane has a unique (up to isomorphism) projective completion, we will talk about **deriving** $\mathcal{P}$ and the **derived plane** $\mathcal{D}(\mathcal{P})$.

Many collineations of $\mathcal{P}$ induce collineations of $\mathcal{D}(\mathcal{P})$. In particular, if $\mathcal{P}$ is a translation plane with respect to $\ell_\infty$, then $\mathcal{D}(\mathcal{P})$ is a translation plane with respect to $\ell'_\infty$. If $\mathcal{P} = \mathrm{PG}(2, q^2)$, then any Baer subline of $\ell_\infty$ is a derivation set and $\mathcal{D}(\mathcal{P})$ is the **Hall plane** $\mathrm{Hall}(q^2)$ of order $q^2$, regardless of the derivation set used. If $D_1, D_2, \ldots, D_k$ are $k$ disjoint Baer sublines of $\ell_\infty$, then we can perform multiple derivations on the Desarguesian plane. If this set of disjoint Baer sublines is "linear" in some well-defined sense (see [71]), then the resulting plane is an **André plane**. See [140] for more details on derivation and non-Desarguesian translation planes.

## 3.3 Spreads

In this section we define spreads in $\mathrm{PG}(3, q)$ and show how to construct a regular spread. We begin with a discussion of reguli in $\mathrm{PG}(3, q)$.

**Definition 3.4.** *A* **regulus** $\mathcal{R}$ *in* $\mathrm{PG}(3, q)$ *is a set of* $q + 1$ *mutually skew lines such that any line which meets three of its lines necessarily meets all* $q + 1$ *of them.*

**Theorem 3.5.** *Any three mutually skew lines in* $\mathrm{PG}(3, q)$ *are contained in a unique regulus in* $\mathrm{PG}(3, q)$.

*Proof.* Suppose $a_1, a_2, a_3$ are three mutually skew lines in $\mathrm{PG}(3, q)$ (that is, they pairwise have no common point). Note that there is a unique line through each point $P \in a_1$ that meets $a_2$ and $a_3$; namely, the transversal line $\langle P, a_2 \rangle \cap \langle P, a_3 \rangle$. Let $V$ be the four-dimensional vector space over $\mathrm{GF}(q)$ that underlies $\mathrm{PG}(3, q)$, so that points are one-dimensional subspaces of $V$, lines are two-dimensional subspaces, and so on. Choose two distinct points, say $\langle e_1 \rangle$ and $\langle e'_1 \rangle$, on $a_1$. Since $V$ is the direct sum of the two-dimensional subspaces underlying $a_2$ and $a_3$, there is a unique point $\langle e_2 \rangle \in a_2$ such that $\langle e_1 + e_2 \rangle \in a_3$. Similarly, there is a unique point $\langle e'_2 \rangle \in a_2$ such that $\langle e'_1 + e'_2 \rangle \in a_3$ (see Figure 3.2). Let $P_t = \langle e_1 + t e'_1 \rangle$ be an arbitrary point on $a_1$, where $t \in \mathrm{GF}(q) \cup \{\infty\}$. Here we are using the convention that $P_\infty = \langle e'_1 \rangle$.

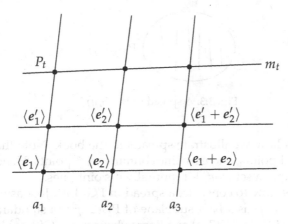

**Fig. 3.2.** Proof of Theorem 3.5

Straightforward computations show that $m_t = \langle e_1 + te_1', e_2 + te_2' \rangle$ is the unique line through $P_t$ that meets each of $a_1, a_2,$ and $a_3$, where again by the usual convention $m_\infty = \langle e_1', e_2' \rangle$. Moreover, $\mathcal{R}' = \{m_t \mid t \in \mathrm{GF}(q) \cup \{\infty\}\}$ is a set of $q+1$ mutually skew lines since $a_1, a_2, a_3$ are mutually skew. In fact, these are the only lines of $\mathrm{PG}(3,q)$ which meet $a_1, a_2,$ and $a_3$ since there is a unique transversal line through each point of $a_1$.

Next let $\ell_s = \langle e_1 + se_2, e_1' + se_2' \rangle$, and let $\mathcal{R} = \{\ell_s \mid s \in \mathrm{GF}(q) \cup \{\infty\}\}$. Then it is similarly seen that $\mathcal{R}$ is a set of $q+1$ mutually skew lines which contains the original lines $a_1, a_2,$ and $a_3$. Straightforward computations show that $\ell_s \cap m_t \neq \emptyset$ for all $s \neq t$.

Now let $m$ be a line which meets any three chosen lines of $\mathcal{R}$. Then there are only $q+1$ choices for $m$, as discussed above. But the $q+1$ lines of $\mathcal{R}'$ meet the three chosen lines of $\mathcal{R}$. Hence $m = m_t$ for some $t$, and thus $m$ meets every line of $\mathcal{R}$. That is, $\mathcal{R}$ is a regulus by definition.

Finally, suppose that $\mathcal{T}$ is another regulus containing $a_1, a_2, a_3$. Since $\mathcal{R}'$ is the unique set of transversal lines to $a_1, a_2,$ and $a_3$, necessarily $\mathcal{R}'$ is the uniquely determined set of transversal lines to $\mathcal{T}$ by the above argument. This implies that $\mathcal{R} = \mathcal{T}$ is the uniquely determined set of transversal lines to $\mathcal{R}'$, and thus $\mathcal{R}$ is the unique regulus containing $a_1, a_2, a_3$.     □

The above proof shows that three mutually skew lines $\ell_1, \ell_2, \ell_3$ are contained in a unique regulus $\mathcal{R} = \{\ell_1, \ell_2, \ell_3, \ldots, \ell_{q+1}\}$. Furthermore, there is a unique regulus $\mathcal{R}' = \{m_1, m_2, \ldots, m_{q+1}\}$ that covers the same points as $\mathcal{R}$, and each line of $\mathcal{R}$ meets each line of $\mathcal{R}'$ in a point. The regulus $\mathcal{R}'$ is called the **opposite** or **reverse** regulus of $\mathcal{R}$. We also note here that a regulus (or its opposite regulus) covers the point set of a hyperbolic quadric in $\mathrm{PG}(3,q)$.

**Definition 3.6.** A **spread** $\mathcal{S}$ in $\mathrm{PG}(3,q)$ *is a set of lines which partitions the points of* $\mathrm{PG}(3,q)$. *A spread* $\mathcal{S}$ *is* **regular** *if given any three lines of* $\mathcal{S}$, *the uniquely determined regulus containing these three lines is contained in* $\mathcal{S}$.

**Fig. 3.3.** A spread in $PG(3, q)$

Figure 3.3 shows how we illustrate spreads in the book. Note that $PG(3, q)$ has $q^3 + q^2 + q + 1$ points, and each line contains $q + 1$ points. Hence a spread of $PG(3, q)$ is simply a set of $q^2 + 1$ mutually disjoint lines.

We now show how to construct a spread in $PG(3, q)$. Just as we can naturally embed $PG(2, q)$ as a Baer subplane of $PG(2, q^2)$ in standard position, we can also embed $\Sigma_q = PG(3, q)$ as a Baer subspace of $PG(3, q^2)$ in standard position. That is, if the points of $PG(3, q^2)$ have homogeneous coordinates $(x_0, x_1, x_2, x_3)$, where $x_0, \ldots, x_3 \in GF(q^2)$, then we embed $\Sigma_q$ so that its points have homogeneous coordinates $(x_0, x_1, x_2, x_3)$, with $x_0, \ldots, x_3 \in GF(q)$.

A line of $\Sigma_q$ naturally extends to a line of $PG(3, q^2)$. If a line of $PG(3, q^2)$ meets $\Sigma_q$ in two points, then it must meet $\Sigma_q$ in $q + 1$ points. Thus a line of $PG(3, q^2)$ meets $\Sigma_q$ in 0, 1, or $q + 1$ points and is called **external**, **tangent**, or **secant** to $\Sigma_q$, respectively. Note also that a plane of $\Sigma_q$ naturally extends to a plane of $PG(3, q^2)$, whereas a plane of $PG(3, q^2)$ meets $\Sigma_q$ in a subspace of $\Sigma_q$ of dimension 1 or 2.

**Lemma 3.7.** *Let $P$ be a point of $PG(3, q^2) \setminus PG(3, q)$. Then $P$ lies on one secant, $q^3 + q^2$ tangents, and $q^4 - q^3$ external lines to $PG(3, q)$.*

*Proof.* If $P$ were on two secants of $\Sigma_q$, then these two secants would generate a plane of $\Sigma_q$ and therefore necessarily meet in $\Sigma_q$, a contradiction. Hence $P$ lies on at most one secant of $\Sigma_q$. Thus there exists a tangent $\ell$ to $\Sigma_q$ through $P$ that meets $\Sigma_q$ in precisely one point, say $Q$. We will show that $\ell$ lies in a unique plane of $\Sigma_q$. There are $q^2 + q + 1$ lines $m_i$ of $\Sigma_q$ through $Q$, as illustrated in Figure 3.4.

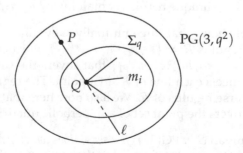

**Fig. 3.4.** Proof of Lemma 3.7

If $\ell$ is in no plane of $\Sigma_q$, then there are $q^2 + q + 1$ distinct planes $\langle m_i, \ell \rangle$ of $\mathrm{PG}(3, q^2)$ that contain $\ell$. This is a contradiction as $\ell$ lies on precisely $q^2 + 1$ planes of $\mathrm{PG}(3, q^2)$. Hence $\ell$ lies in at least one plane of $\Sigma_q$. However, $\ell$ cannot lie in two planes of $\Sigma_q$ as $\ell$ is not a line of $\Sigma_q$. Therefore $\ell$ lies in a unique plane of $\Sigma_q$, and we denote this plane by $\pi$. As $P$ lies on at most one secant of $\Sigma_q$, $P$ lies on at least $q^3 + q^2$ tangents of $\Sigma_q$, and thus there is a line $\ell'$ through $P$ tangent to $\Sigma_q$ that is not in $\pi$. Let $\pi'$ be the unique plane of $\Sigma_q$ containing $\ell'$. Then $P$ is on the line $\pi \cap \pi'$ of $\Sigma_q$; that is, $P$ lies on at least one line of $\Sigma_q$. Hence $P$ lies on exactly one secant of $\Sigma_q$. As the $q^4 + q^2 + 1$ lines of $\mathrm{PG}(3, q^2)$ through $P$ partition the $q^3 + q^2 + q + 1$ points of $\Sigma_q$, we see that $P$ lies on one secant to $\Sigma_q$, $q^3 + q^2$ tangents, and $q^4 - q^3$ external lines.   $\square$

**Theorem 3.8.** Let $\ell = \{P_1, P_2, \ldots, P_{q^2+1}\}$ be a line of $\mathrm{PG}(3, q^2)$ disjoint from $\Sigma_q = \mathrm{PG}(3, q)$. For each point $P_i \in \ell$, $i = 1, 2, \ldots, q^2 + 1$, let $m_i$ be the unique secant of $\mathrm{PG}(3, q)$ through $P_i$. Then $m_1, m_2, \ldots, m_{q^2+1}$ meet $\mathrm{PG}(3, q)$ in a spread of $\mathrm{PG}(3, q)$ (see Figure 3.5).

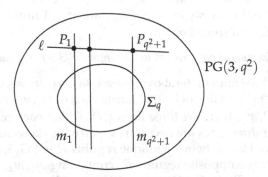

**Fig. 3.5.** Construction of a spread in $\mathrm{PG}(3, q)$

*Proof.* The lines $\ell, m_1, m_2, \ldots, m_{q^2+1}$ exist by Lemma 3.7. If $m_1$ and $m_2$ meet in a point, then they lie in a plane $\pi$ of $\Sigma_q$. Hence in $\mathrm{PG}(3, q^2)$, $\pi$ is a projective plane of order $q^2$ and $\ell$ is a line of $\pi$ disjoint from $\pi \cap \Sigma_q$, a Baer subplane of $\pi$. However, by Theorem 3.2, a Baer subplane has no external lines. Hence $m_1, m_2, \ldots, m_{q^2+1}$ are $q^2 + 1$ mutually skew lines which necessarily intersect $\Sigma_q$ in a spread of $\Sigma_q$.   $\square$

This spread also can be constructed by using the conjugation map of $\mathrm{GF}(q^2)$:

$$\sigma : x \longmapsto x^q.$$

As previously discussed, this field automorphism induces an automorphic collineation of $\mathrm{PG}(3, q^2)$: if $P = (x_0, x_1, x_2, x_3)$ is a point of $\mathrm{PG}(3, q^2)$, then

$P^\sigma = (x_0^q, x_1^q, x_2^q, x_3^q)$. Hence $\sigma$ fixes $\Sigma_q = \mathrm{PG}(3,q)$ pointwise, and these are the only points fixed by $\sigma$. If $\ell$ is a line of $\mathrm{PG}(3,q^2)$ that meets $\Sigma_q$ in $q+1$ points, then $\ell$ is fixed by $\sigma$. Conversely, let $m$ be a line of $\mathrm{PG}(3,q^2)$ that is fixed by $\sigma$. Let $P \in m$ with $P \notin \Sigma_q$. Then $P^\sigma \neq P$ and $P^\sigma \in m$. Hence $m = PP^\sigma$ is the unique fixed line through $P$. Now by Lemma 3.7, $P$ lies on exactly one secant of $\Sigma_q$, and this secant is fixed by $\sigma$. Thus $m$ is the unique secant of $\Sigma_q$ through $P$. That is, if $m$ is fixed by $\sigma$, then $m$ is secant to $\Sigma_q$, and the lines fixed by $\sigma$ are precisely the lines secant to $\Sigma_q$. We now use this fact to give a different construction of the spread obtained in Theorem 3.8.

**Theorem 3.9.** *Let* $\ell = \{P_1, P_2, \dots, P_{q^2+1}\}$ *be a line disjoint from* $\mathrm{PG}(3,q)$. *Let* $m_i = P_i P_i^\sigma$, *for* $i = 1, 2, \dots, q^2 + 1$. *Then* $m_1, m_2, \dots, m_{q^2+1}$ *meet* $\mathrm{PG}(3,q)$ *in a spread of* $\mathrm{PG}(3,q)$.

*Proof.* If $m_i = P_i P_i^\sigma$, then $m_i^\sigma = P_i^\sigma (P_i^\sigma)^\sigma = P_i^\sigma P_i = m_i$ and $m_i$ is fixed by $\sigma$. Hence by the above argument, $m_i$ meets $\Sigma_q$ in $q+1$ points. Now suppose that $m_i$ and $m_j$ meet in a point for some $i \neq j$. Then $m_i$ and $m_j$ span a plane which contains both the line $\ell$ and the line $\ell^\sigma = \{P_1^\sigma, P_2^\sigma, \dots, P_{q^2+1}^\sigma\}$. Hence $\ell$ and $\ell^\sigma$ meet in some point $P$, and $P^\sigma = (\ell \cap \ell^\sigma)^\sigma = \ell^\sigma \cap \ell = P$. Thus $P$ is in $\Sigma_q$, a contradiction. Hence $m_1, m_2, \dots, m_{q^2+1}$ are $q^2 + 1$ mutually skew lines which intersect $\Sigma_q$ in a spread of $\Sigma_q$. □

**Theorem 3.10.** *The spread constructed in Theorem 3.8/3.9 is regular.*

*Proof.* Take any three lines of the above spread, say $m_1, m_2, m_3$ without loss of generality, and let $\ell'$ be a line of $\Sigma_q$ that meets each of $m_1, m_2, m_3$ in a point of $\Sigma_q$. Then in $\mathrm{PG}(3,q^2)$, there are three lines $\ell, \ell^\sigma, \ell'$ that each meet $m_1, m_2, m_3$ in a point. These three lines are necessarily pairwise skew since $m_1, m_2, m_3$ are pairwise skew. Hence there is a unique regulus $\mathcal{R}$ in $\mathrm{PG}(3,q^2)$ containing $\ell, \ell^\sigma, \ell'$ with a unique opposite regulus $\mathcal{R}'$ containing $m_1, m_2, m_3$. Moreover, the unique regulus $\mathcal{R}_0$ in $\Sigma_q$ determined by $m_1 \cap \Sigma_q$, $m_2 \cap \Sigma_q$, $m_3 \cap \Sigma_q$ is contained in $\mathcal{R}'$ (when the lines are extended over $\mathrm{GF}(q^2)$). Since these (extended) lines of $\mathcal{R}_0$ are secants of $\Sigma_q$ which meet the line $\ell$, they must be lines of the spread constructed in Theorem 3.8. Hence this spread is regular by definition. □

Conversely, every regular spread can be constructed in this way. See Bruck [71, Theorem 5.3] for a proof of this converse result.

Hence, if $\mathcal{S}$ is a regular spread in $\mathrm{PG}(3,q)$, then associated with $\mathcal{S}$ are two unique lines $\ell, \ell^\sigma$ in $\mathrm{PG}(3,q^2) \setminus \mathrm{PG}(3,q)$ called the **transversals** of $\mathcal{S}$. Either of these transversals uniquely determines $\mathcal{S}$, and conversely $\mathcal{S}$ uniquely determines its two transversals. In particular, note that $\ell, \ell^\sigma$ are the only lines of $\mathrm{PG}(3,q^2)$ that are disjoint from $\mathrm{PG}(3,q)$ and that meet every (extended) line of $\mathcal{S}$ (see [71]).

We close this section with one last observation about regular spreads in $\mathrm{PG}(3,q)$.

**Theorem 3.11.** *Let $S$ be a regular spread in $PG(3, q)$, and let $m$ be a line of $PG(3, q)$ which is not in $S$. Then the $q + 1$ lines of $S$ which meet $m$ necessarily form a regulus of $S$.*

*Proof.* Since the lines of $S$ partition the points of $PG(3, q)$ and $m$ is not a line of $S$, we see that precisely $q + 1$ lines of $S$ meet $m$ (in one point each). Let $\ell_1, \ell_2, \ell_3, \ldots, \ell_{q+1}$ be these lines of $S$, and let $\mathcal{R}$ be the unique regulus of $PG(3, q)$ determined by $\ell_1, \ell_2, \ell_3$. Now $m$ meets each of $\ell_1, \ell_2, \ell_3$ and hence $m$ meets each line of $\mathcal{R}$. That is, $m$ is a line of the opposite regulus $\mathcal{R}'$. Since the spread $S$ is regular, the regulus $\mathcal{R}$ is a subset of $S$. Therefore the uniquely determined $q + 1$ lines of $S$ that meet $m \in \mathcal{R}'$ must necessarily be the lines of $\mathcal{R}$, proving the result. $\qquad\square$

## 3.4 The Bruck-Bose Representation

In this section we introduce the linear representation of a finite translation plane $\mathcal{P}$ of dimension at most two over its kernel, an idea which was developed independently by André [12], Segre [197], and Bruck and Bose [73, 74]. André presented the representation as a group theoretic construction, whereas Bruck and Bose used a vector space approach. We follow the approach used by Bruck and Bose.

### 3.4.1 The Bruck-Bose Construction

Let $\Sigma_\infty$ be a hyperplane of $PG(4, q)$ and let $S$ be a spread of $\Sigma_\infty$. We refer to a plane of $PG(4, q)$ that is not contained in $\Sigma_\infty$ as a plane of $PG(4, q) \backslash \Sigma_\infty$. Consider the following incidence structure $\mathcal{A}(S)$:

- the *points* of $\mathcal{A}(S)$ are the points of $PG(4, q) \backslash \Sigma_\infty$;
- the *lines* of $\mathcal{A}(S)$ are the planes of $PG(4, q) \backslash \Sigma_\infty$ that contain an element of $S$;
- *incidence* in $\mathcal{A}(S)$ is induced by incidence in $PG(4, q)$.

Figure 3.6 illustrates this construction.

**Theorem 3.12.** *The incidence structure $\mathcal{A}(S)$ is an affine plane of order $q^2$.*

*Proof.* We begin with some counting. The number of points in $\mathcal{A}(S)$ is equal to the number of points in $PG(4, q)$ minus the number of points in $PG(3, q)$; that is, $(q^4 + q^3 + q^2 + q + 1) - (q^3 + q^2 + q + 1) = q^4$. The number of points on a line of $\mathcal{A}(S)$ is the number of points of $PG(4, q) \backslash \Sigma_\infty$ that lie in a plane that meets $\Sigma_\infty$ in a line; that is, $(q^2 + q + 1) - (q + 1) = q^2$.

Let $P, Q$ be two points in $\mathcal{A}(S)$, then they correspond to two points in $PG(4, q) \backslash \Sigma_\infty$ which we also denote by $P, Q$. In $PG(4, q)$ there is a unique line

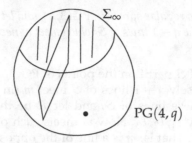

**Fig. 3.6.** The Bruck-Bose construction

through $P$ and $Q$ which necessarily meets $\Sigma_\infty$ in a point. The point $PQ \cap \Sigma_\infty$ lies on exactly one line $\ell$ of the spread $\mathcal{S}$. Thus $\langle \ell, PQ \rangle$ is the unique plane of $PG(4,q) \backslash \Sigma_\infty$ containing $P, Q$ and a spread line. Hence in $\mathcal{A}(\mathcal{S})$ there is a unique line through $P$ and $Q$. Thus $\mathcal{A}(\mathcal{S})$ is a 2-$(q^4, q^2, 1)$ design, and hence an affine plane of order $q^2$. □

The above construction is referred to as the **Bruck-Bose construction**, or **Bruck-Bose representation**, of $\mathcal{A}(\mathcal{S})$ in $PG(4,q)$. The affine plane $\mathcal{A}(\mathcal{S})$ can be uniquely completed to a projective plane $\mathcal{P}(\mathcal{S})$ by adding a line at infinity, $\ell_\infty$. Planes of $PG(4,q) \backslash \Sigma_\infty$ through a given element $t$ of $\mathcal{S}$ correspond to lines of $\mathcal{A}(\mathcal{S})$ in the same parallel class. In the completion $\mathcal{P}(\mathcal{S})$ the lines in this parallel class meet in the point $T$ of $\ell_\infty$ that corresponds to the line $t$ of $\mathcal{S}$ (see Figure 3.7). Hence the points of $\ell_\infty$ are in one-to-one correspondence with the lines of the spread $\mathcal{S}$.

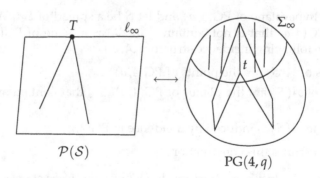

**Fig. 3.7.** The projective Bruck-Bose construction

**Notation Summary:** In summary, we use the following notation for the Bruck-Bose correspondence between $\mathcal{P}(\mathcal{S})$ and $PG(4,q)$. If $P$ is a point of $\mathcal{A}(\mathcal{S})$, then we also denote the corresponding point of $PG(4,q) \backslash \Sigma_\infty$ by $P$. The points of $\ell_\infty$ correspond to lines of the spread $\mathcal{S}$ in $\Sigma_\infty$. If $T$ is a point of $\ell_\infty$, then we denote the corresponding spread line by $t$. We also use the phrase *a subspace of* $PG(4, q) \setminus \Sigma_\infty$ to refer to a subspace of $PG(4, q)$ that is not contained in $\Sigma_\infty$. In Section 4.1 we study unitals in the Bruck-Bose representation. If $U$ is a unital of $\mathcal{P}(S)$, then we denote the corresponding point set in $PG(4, q)$ by $\mathcal{U}$.

Bruck and Bose [73] showed that $\mathcal{P}(S)$ is a translation plane of dimension at most two over its kernel, and conversely, any such translation plane can be constructed from a spread of $PG(3, q)$ in this way. Further, $\mathcal{P}(S) \cong PG(2, q^2)$ if and only if the spread $S$ is regular.

The Bruck-Bose representation has proved to be very useful for studying objects such as unitals in $PG(2, q^2)$ and other translation planes. Section 3.4.4 develops a simple way of relating coordinates of points and lines in $\mathcal{P}(S)$ to subspaces in $PG(4, q)$. The Bruck-Bose construction also generalizes to represent $PG(2, q^n)$ in $PG(2n, q)$. See [36] for a treatment focusing on the representation of unitals and Baer subplanes of $PG(2, q^n)$ in $PG(2n, q)$.

### 3.4.2 Baer Subplanes and Baer Sublines in Bruck-Bose

In this section we determine the representation of Baer subplanes and Baer sublines of $\mathcal{P}(S)$ in $PG(4, q)$. We have four structures to investigate: Baer subplanes secant to $\ell_\infty$, Baer subplanes tangent to $\ell_\infty$, Baer sublines with a point on $\ell_\infty$, and Baer sublines disjoint from $\ell_\infty$. Recall that if $\mathcal{B}$ is a Baer subplane of $\mathcal{P}(S)$ secant to $\ell_\infty$, then $\mathcal{B} \setminus \ell_\infty$ is an affine plane which we call an affine Baer subplane.

Let $\pi$ be a plane of $PG(4, q)$ that does not contain an element of $S$. A simple counting argument shows that every plane in $\Sigma_\infty \cong PG(3, q)$ contains a unique line of $S$, so necessarily $\pi \not\subseteq \Sigma_\infty$. Hence $\pi$ must meet $\Sigma_\infty$ in a line $m$. Since $m$ is not in the spread, $m$ must meet $q + 1$ lines of $S$. We denote this set of $q + 1$ lines by $\mathcal{R}$ (see Figure 3.8). Note that if $S$ is a regular spread, then $\mathcal{R}$ must be a regulus by Theorem 3.11.

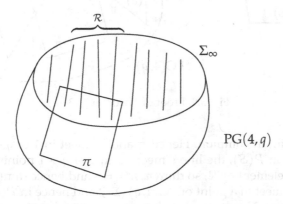

**Fig. 3.8.** A plane of $PG(4, q)$ not containing a spread line

In $\mathcal{P}(\mathcal{S})$, $\pi$ corresponds to a set $\mathcal{B}$ consisting of $q^2$ affine points and $q+1$ points of $\ell_\infty$ (corresponding to the $q+1$ lines in $\mathcal{R}$). We now show that $\mathcal{B}$ is a Baer subplane of $\mathcal{P}(\mathcal{S})$. Moreover, if $\mathcal{P}(\mathcal{S})$ is Desarguesian, we show that all Baer subplanes secant to $\ell_\infty$ (that is, meeting $\ell_\infty$ in $q+1$ points) arise in this way.

**Theorem 3.13.** *Consider the translation plane $\mathcal{P}(\mathcal{S})$ defined above.*

1. *A plane of $\mathrm{PG}(4,q)$ that does not contain a line of $\mathcal{S}$ represents a Baer subplane in $\mathcal{P}(\mathcal{S})$ which is secant to $\ell_\infty$.*
2. *If $\mathcal{P}(\mathcal{S}) \cong \mathrm{PG}(2,q^2)$, then every Baer subplane of $\mathcal{P}(\mathcal{S})$ which is secant to $\ell_\infty$ is represented by a plane of $\mathrm{PG}(4,q)$ that does not contain a spread line.*

*Proof.* To prove (1), let $\pi$ be a plane of $\mathrm{PG}(4,q)$ that does not contain a line of $\mathcal{S}$. As discussed above, the points in $\pi$ correspond in $\mathcal{P}(\mathcal{S})$ to a set $\mathcal{B}$ of $q^2 + q + 1$ points: $q^2$ affine points (corresponding to the points in $\pi \setminus \Sigma_\infty$) and $q+1$ points on $\ell_\infty$ (corresponding to the lines $\mathcal{R}$ in $\mathcal{S}$). We define a line of $\mathcal{B}$ to be the intersection of a line of $\mathcal{P}(\mathcal{S})$ with $\mathcal{B}$, provided this intersection has size greater than 1. We first show that every line of $\mathcal{B}$ has $q+1$ points. If $\ell$ is a line of $\mathcal{P}(\mathcal{S})$ which is not equal to $\ell_\infty$, then in $\mathrm{PG}(4,q)$ the line $\ell$ corresponds to a plane $\pi'$ that contains a line of the spread. There are two cases: either (a) $\pi'$ meets the line $m = \pi \cap \Sigma_\infty$, or (b) $\pi'$ does not meet $m$. These cases are shown in Figure 3.9. In case (a), $\pi'$ contains a line of $\mathcal{R}$, so $\pi$ and $\pi'$ have

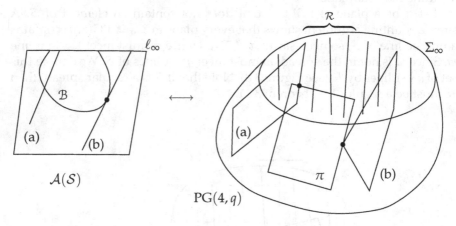

**Fig. 3.9.** Proof of Theorem 3.13

at least one point in common. Hence $\pi$ and $\pi'$ meet in 1 or $q+1$ points in this case. Thus in $\mathcal{P}(\mathcal{S})$, the line $\ell$ meets $\mathcal{B}$ in 1 or $q+1$ points. In case (b), $\pi'$ contains no element of $\mathcal{R}$, so $\dim\langle \pi, \pi' \rangle = 4$ and hence $\dim(\pi \cap \pi') = 0$. Thus $\pi$ and $\pi'$ meet in a point of $\mathrm{PG}(4,q) \setminus \Sigma_\infty$, and hence in $\mathcal{P}(\mathcal{S})$, the line $\ell$ meets $\mathcal{B}$ in 1 point. Therefore every line of $\mathcal{B}$ contains $q+1$ points.

Now consider two points $P, Q$ in $\mathcal{B}$. These points lie on a unique line $\ell$ of $\mathcal{P}(\mathcal{S})$. Thus $|\ell \cap \mathcal{B}| \geq 2$ and hence $\ell$ meets $\mathcal{B}$ in $q + 1$ points. That is, $P$ and $Q$ lie on a unique line of $\mathcal{B}$. Hence the points and lines of $\mathcal{B}$ form a $2\text{-}(q^2 + q + 1, q + 1, 1)$ design, and therefore a projective plane of order $q$. In particular, $\mathcal{B}$ is a Baer subplane of $\mathcal{P}(\mathcal{S})$.

To prove (2), we assume that $\mathcal{P}(\mathcal{S}) \cong \mathrm{PG}(2, q^2)$. Part (1) shows that a plane of $\mathrm{PG}(4, q)$ which does not contain a spread line gives rise to a Baer subplane of $\mathrm{PG}(2, q^2)$ that is secant to $\ell_\infty$. We now show that all such Baer subplanes arise in this way by showing that the two sets have the same size. The number of lines in $\Sigma_\infty$ not in $\mathcal{S}$ is

$$(q^2 + 1)(q^2 + q + 1) - (q^2 + 1) = (q^2 + 1)(q^2 + q).$$

Through each of these lines there are $q^2 + q + 1$ planes in $\mathrm{PG}(4, q)$, of which $q + 1$ are in $\Sigma_\infty$. Hence there are $(q^2 + 1)(q^2 + q)q^2$ planes of $\mathrm{PG}(4, q) \backslash \Sigma_\infty$ that contain no spread line. To count the number of Baer subplanes secant to $\ell_\infty$, we use Theorem 2.8. Namely, in $\mathrm{PG}(2, q^2)$ there is a unique Baer subplane through each quadrangle. Hence there is a unique Baer subplane secant to $\ell_\infty$ through any quadrangle which contains two points of $\ell_\infty$. The number of such quadrangles is $(q^2 + 1)(q^2)(q^4)(q^4 - 2q^2 + 1)/4!$. The resulting Baer subplanes are not necessarily distinct as each such Baer subplane $\mathcal{B}$ can be determined by any quadrangle in $\mathcal{B}$ with two points on $\ell_\infty$. Hence the number of distinct Baer subplanes secant to $\ell_\infty$ is

$$\frac{(q^2 + 1)(q^2)(q^4)(q^4 - 2q^2 + 1)/4!}{(q + 1)(q)(q^2)(q^2 - 2q + 1)/4!} = (q^2 + 1)q^3(q + 1).$$

This is the same number as the number of planes of $\mathrm{PG}(4, q) \backslash \Sigma_\infty$ that do not contain a spread line, proving the result. $\qquad\square$

The following corollary of Theorem 3.13 is immediate.

**Corollary 3.14.** *Consider the translation plane $\mathcal{P}(\mathcal{S})$ defined above.*

1. *Lines of $\mathrm{PG}(4, q) \backslash \Sigma_\infty$ that meet $\Sigma_\infty$ in a point on the line $t$ of $\mathcal{S}$ represent Baer sublines of $\mathcal{P}(\mathcal{S})$ that meet $\ell_\infty$ in the corresponding point $T$.*
2. *If $\mathcal{P}(\mathcal{S}) \cong \mathrm{PG}(2, q^2)$, then every Baer subline of $\mathcal{P}(\mathcal{S})$ which meets $\ell_\infty$ in a point is represented by a line of $\mathrm{PG}(4, q) \backslash \Sigma_\infty$.*

For Baer sublines contained in $\ell_\infty$, we have the following result.

**Corollary 3.15.** *Consider the translation plane $\mathcal{P}(\mathcal{S})$ defined above.*

1. *A regulus of $\mathcal{S}$, if any such exist, represents a Baer subline of $\mathcal{P}(\mathcal{S})$ contained in $\ell_\infty$.*
2. *If $\mathcal{P}(\mathcal{S}) \cong \mathrm{PG}(2, q^2)$, then every Baer subline of $\mathcal{P}(\mathcal{S})$ contained in $\ell_\infty$ is represented by a regulus of $\mathcal{S}$.*

*Proof.* Part (1) follows directly from Theorem 3.13. Note that there will be many Baer subplanes of $\mathcal{P}(\mathcal{S})$ which contain this Baer subline. To prove (2), assume that $\mathcal{S}$ is regular and hence $\mathcal{P}(\mathcal{S}) \cong \mathrm{PG}(2, q^2)$. Then Theorem 3.11 implies that any plane of $\mathrm{PG}(4, q)$ not containing a spread line determines a regulus in $\mathcal{S}$. Furthermore, the number of reguli in $\mathcal{S}$ is equal to the number of Baer sublines in $\ell_\infty$ (since any three lines lie in a unique regulus and any three points lie in a unique Baer subline). Therefore the correspondence in part (1) is a bijection for the Desarguesian plane $\mathrm{PG}(2, q^2)$. $\qquad\square$

It must be noted that Theorem 3.13(2) does not hold in general for non-Desarguesian translation planes of dimension two over their kernel. Vincenti [216] and independently Freeman [119] showed that certain two-dimensional translation planes of order $q^4$ contain affine Baer subplanes that are not represented by planes of $\mathrm{PG}(4, q^2)$. In the survey article [219], Vincenti quotes the following representations of affine Baer subplanes in non-Desarguesian planes.

**Theorem 3.16 ([216]).** *Let $\mathcal{P}(\mathcal{S})$ be a non-Desarguesian plane of order $q^4$ of dimension at most two over its kernel, so that $\pi$ can be represented in $\mathrm{PG}(4, q^2)$ using the Bruck-Bose representation. Thus $\mathcal{S}$ is a spread in the hyperplane $\Sigma_\infty \cong \mathrm{PG}(3, q^2)$ of $\mathrm{PG}(4, q^2)$. Then the following represent Desarguesian affine Baer subplanes of $\mathcal{P}(\mathcal{S})$:*

- *planes of $\mathrm{PG}(4, q^2)$,*
- *four-dimensional subgeometries $\Gamma = \mathrm{PG}(4, q)$ meeting the spread $\mathcal{S}$ in a regular spread of $\Gamma \cap \Sigma_\infty \cong \mathrm{PG}(3, q)$.*

*Moreover, four-dimensional subgeometries $\mathrm{PG}(4, q)$ that meet the spread $\mathcal{S}$ in a non-regular spread of $\mathrm{PG}(4, q) \cap \Sigma_\infty \cong \mathrm{PG}(3, q)$ represent non-Desarguesian affine Baer subplanes of $\mathcal{P}(\mathcal{S})$.*

Vincenti's survey notes that this list is not exhaustive and cites [53, Theorem 3] for a complete list of all possibilities.

We have yet to determine the representation in $\mathrm{PG}(4, q)$ for Baer subplanes of $\mathcal{P}(\mathcal{S})$ that are tangent to $\ell_\infty$ and for Baer sublines that are disjoint from $\ell_\infty$. In general, it is difficult to describe structures in $\mathrm{PG}(4, q)$ that represent such objects in $\mathcal{P}(\mathcal{S})$. However, when $\mathcal{P}(\mathcal{S}) \cong \mathrm{PG}(2, q^2)$, we can completely describe how such objects are represented.

The next theorem shows that a Baer subline of $\mathrm{PG}(2, q^2)$ which has no point on $\ell_\infty$ is represented in $\mathrm{PG}(4, q)$ by a conic in a plane passing through a spread line. This conic is irreducible and is disjoint from $\Sigma_\infty$; that is, it is an "ellipse." However, we show below that not every irreducible conic disjoint from $\Sigma_\infty$ and contained in a plane about a spread line represents a Baer subline. Those conics $\mathcal{C}$ in $\mathrm{PG}(4, q) \backslash \Sigma_\infty$ that represent a Baer subline of $\mathrm{PG}(2, q^2)$ disjoint from $\ell_\infty$ will be called **Baer conics**.

**Theorem 3.17.** *Every Baer subline of* $\mathcal{P}(\mathcal{S}) \cong \mathrm{PG}(2, q^2)$ *containing no point of* $\ell_\infty$ *is represented by an irreducible conic in some plane of* $\mathrm{PG}(4, q) \setminus \Sigma_\infty$ *passing through a line of the regular spread* $\mathcal{S}$.

A standard approach to proving this theorem is through the use of inversive planes (see [39]), combinatorial objects that will be defined in Chapter 6. We will present a direct proof of this result through the use of coordinates, once they have been properly introduced in the Bruck-Bose model in Section 3.4.4; see page 56.

First we present the following result of Metz [172].

**Theorem 3.18.** *Let* $\pi$ *be a plane of* $\mathrm{PG}(4, q) \setminus \Sigma_\infty$ *passing through a line* $t$ *of the regular spread* $\mathcal{S}$. *Let* $\ell$ *be the corresponding line of* $\mathcal{P}(\mathcal{S}) \cong \mathrm{PG}(2, q^2)$ *which meets* $\ell_\infty$ *in the point* $T$. *If* $q > 2$, *not every conic in* $\pi \setminus t$ *represents a Baer subline of* $\ell \setminus T$.

*Proof.* We show that for $q > 2$, the number of irreducible conics contained in $\pi \setminus t$ is greater than the number of Baer sublines contained in $\ell \setminus T$. By Theorem 2.6, the number of Baer sublines of $\ell \cong \mathrm{PG}(1, q^2)$ is

$$\binom{q^2 + 1}{3} \Big/ \binom{q + 1}{3} = q(q^2 + 1),$$

of which $q(q + 1)$ pass through the point $T$. Hence the number of Baer sublines in $\ell \setminus T$ is $q^2(q - 1)$. To count the number $x$ of irreducible conics in $\pi \setminus t$, we first count in two ways the ordered pairs $(\mathcal{C}, m)$, where $m$ is a line external to some irreducible conic $\mathcal{C}$ in $\pi$. On the one hand, there are $q^5 - q^2$ irreducible conics in $\pi$ (see [131]), and any such conic has $q(q - 1)/2$ lines of $\pi$ external to it. On the other hand, there are $q^2 + q + 1$ lines in $\pi$, and each of these lines has the same number $x$ of irreducible conics disjoint from it. Thus the number $x$ of irreducible conics in $\pi$ disjoint from a given line, such as $t$, is $q^3(q - 1)^2/2$. Hence if $q > 2$, there are more irreducible conics in $\pi \setminus t$ than Baer sublines contained in $\ell \setminus T$. □

We now consider Baer subplanes of $\mathrm{PG}(2, q^2)$ tangent to $\ell_\infty$; that is, **non-affine Baer subplanes**. It turns out that they correspond to *ruled cubic surfaces* in $\mathrm{PG}(4, q)$. We begin by considering the following set of points in $\mathrm{PG}(4, q)$. Let $\mathcal{C}$ be a conic in a plane $\pi$, and let $t$ be a line in $\mathrm{PG}(4, q)$ disjoint from $\pi$. Consider a projectivity (homography) $\phi$ of $\mathrm{PG}(4, q)$ that maps the points of $t$ to the points of $\mathcal{C}$. Labeling the points on the line $t$ by $V_1, V_2, \ldots, V_{q+1}$, let $\mathcal{B}$ be the set of $(q + 1)^2$ points on the $q + 1$ lines $\ell_i = V_i V_i^\phi$, for $i = 1, 2, \ldots, q + 1$, see Figure 3.10.

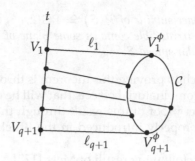

**Fig. 3.10.** A ruled cubic surface of $PG(4, q)$

Bernasconi and Vincenti [41] proved the following results about this set $\mathcal{B}$, whose proofs rely on the theory of varieties and hence are omitted here. The interested reader is encouraged to see [131] for the standard terminology of algebraic varieties.

**Theorem 3.19.** *Using the above notation, the following results hold.*

1. *The points of $\mathcal{B}$ form a rational ruled variety $V_2^3$ of dimension two and order three. We call $\mathcal{B}$ a **ruled cubic surface**. The line $t$ is called the **line directrix** of $\mathcal{B}$, and the lines $\ell_1, \ell_2, \ldots, \ell_{q+1}$ are called **generators** of $\mathcal{B}$.*
2. *$\mathcal{B}$ has a unique line directrix and a unique system of generators.*
3. *The generators are mutually skew, and no three distinct generators lie in a hyperplane of $PG(4, q)$.*
4. *A hyperplane of $PG(4, q)$ meets $\mathcal{B}$ in a cubic curve (this curve may be reducible or have some component in an extension of $PG(4, q)$).*
5. *If $P$ and $Q$ are two points of $\mathcal{B} \setminus t$ on distinct generators, then there is a unique conic of $\mathcal{B}$ through $P$ and $Q$. Hence*
   (a) *$\mathcal{B}$ contains precisely $q^2$ conics, called **conic directrices**.*
   (b) *Each conic directrix is disjoint from $t$ and contains a unique point on each generator of $\mathcal{B}$.*
   (c) *Two distinct conic directrices meet in a unique point of $\mathcal{B}$.*

This article also refers to Bertini [42] for a proof that any two ruled cubic surfaces in $PG(4, q)$ are projectively equivalent.

The next theorem gives the correspondence between nonaffine Baer subplanes of $PG(2, q^2)$ and ruled cubic surfaces. It was proved independently by Vincenti [217] and Bose, Freeman, and Glynn [62] (in the latter reference the surface is called a **twisted ladder**). A direct geometric proof is given by Quinn and Casse [183], although their proof still uses the theory of algebraic varieties. We state the fundamental correspondence here without proof.

**Theorem 3.20.** *Let $\mathcal{B}$ be a Baer subplane in $PG(2, q^2)$ tangent to $\ell_\infty$ at the point $T$. Then $\mathcal{B}$ corresponds to a ruled cubic surface $\mathcal{B}$ in $PG(4, q)$ with line directrix $t$.*

An immediate question to ask is whether the converse of this theorem is true in $\mathrm{PG}(2, q^2)$. By Theorem 3.18, there exist conics in $\mathrm{PG}(4, q) \backslash \Sigma_\infty$ which do not represent Baer sublines. Hence it follows that there exist ruled cubic surfaces which do not represent Baer subplanes and yet have a line directrix which is in the spread. Ruled cubic surfaces which do represent Baer subplanes are called **Baer ruled cubics**. Now suppose that we have a ruled cubic surface $\mathcal{B}$ that has a spread line $t$ for its line directrix, and suppose $\mathcal{B}$ contains a Baer conic. Does $\mathcal{B}$ necessarily correspond to a Baer subplane of $\mathrm{PG}(2, q^2)$? The next theorem shows that the answer to this question is no; there are more ruled cubic surfaces in $\mathrm{PG}(4, q)$ containing a given spread line and a given Baer conic than there are Baer subplanes of $\mathrm{PG}(2, q^2)$ tangent to $\ell_\infty$ at a given point and containing a given Baer subline disjoint from $\ell_\infty$.

**Theorem 3.21.** *In the Bruck-Bose representation of* $\mathrm{PG}(2, q^2)$ *in* $\mathrm{PG}(4, q)$*, let $t$ be a spread line, and let $\mathcal{C}$ be a Baer conic in a plane of* $\mathrm{PG}(4, q) \backslash \Sigma_\infty$ *about some spread line distinct from $t$. Then $t$ and $\mathcal{C}$ are contained in* $(q + 1)(q)(q - 1)$ *ruled cubic surfaces, $q + 1$ of which are Baer ruled cubics.*

*Proof.* In $\mathrm{PG}(2, q^2)$, the spread line $t$ corresponds to a point $T$ of $\ell_\infty$. The Baer conic $\mathcal{C}$ corresponds to a Baer subline $c$ (disjoint from $\ell_\infty$) on a line not through $T$. Let $\ell$ be a line of $\mathrm{PG}(2, q^2)$ joining $T$ to a point on $c$. If we pick a third point $P$ on $\ell$, then $P, T$ and $\ell \cap c$ lie in a unique Baer subline $b$ (by Theorem 2.6). The two intersecting Baer sublines $b$ and $c$ contain a quadrangle ($T$, $P$, and any two points of $c \backslash \ell$), hence by Theorem 2.8 they lie in a unique Baer subplane. This Baer subplane must contain $c \cap \ell$, and so contains the Baer sublines $c$ and $b$. Note that this Baer subplane is tangent to $\ell_\infty$ as $c$ does not meet $\ell_\infty$. The points $T$ and $c \cap \ell$ are contained in

$$(q^2 - 1)/(q - 1) = q + 1$$

Baer sublines of $\ell$, hence $T$ and $c$ are contained in $q + 1$ Baer subplanes (which are all tangent to $\ell_\infty$). Therefore in $\mathrm{PG}(4, q)$, $t$ and $\mathcal{C}$ are contained in $q + 1$ Baer ruled cubics. If $Q$ is a point on $\mathcal{C}$ and the points of $t$ are labeled $V_1, V_2, \ldots, V_{q+1}$, then there is a unique Baer ruled cubic containing $t, \mathcal{C}$, and the line $QV_i$. That is, one generator uniquely determines the Baer ruled cubic.

We now investigate how many ruled cubic surfaces contain $t$ and $\mathcal{C}$. To do this, we use the representation of ruled cubic surfaces based on the projectivities previously discussed. A projectivity is uniquely determined by the images of three distinct points (see [131, Section 6.1]), so there are $(q + 1)(q)(q - 1)$ such projectivities. Let $\phi$ and $\phi'$ be two distinct projectivities of $\mathrm{PG}(4, q)$ mapping $t$ to $\mathcal{C}$ which give rise to the two ruled cubic surfaces

$$\mathcal{B} = \bigcup_{i=1}^{q+1} V_i V_i^\phi \quad \text{and} \quad \mathcal{B}' = \bigcup_{i=1}^{q+1} V_i V_i^{\phi'}.$$

As $\phi \neq \phi'$, there is at least one point $V \in t$ with $V^\phi \neq V^{\phi'}$. Label the generator of $\mathcal{B}$ through $V$ by $\ell$ and the generator of $\mathcal{B}'$ through $V$ by $\ell'$. If $\mathcal{B} = \mathcal{B}'$, then either $B$ has two generators through $V$ or $B$ has two systems of generators. Neither of these is possible by Theorem 3.19. Hence $B \neq B'$. Thus there are $(q + 1)(q)(q - 1)$ distinct ruled cubic surfaces containing $t$ and $C$.    □

We note that if $\mathcal{P}(S)$ is a non-Desarguesian translation plane, then a ruled cubic surface in $PG(4, q)$ does not give rise to a Baer subplane of $\mathcal{P}(S)$. That is, if a ruled cubic surface corresponds to a Baer subplane of $\mathcal{P}(S)$, then $S$ is regular and $\mathcal{P}(S) \cong PG(2, q^2)$. This was first proved by Bernasconi and Vincenti [41]. Here we give the shorter proof given by Casse and Quinn [89].

**Theorem 3.22.** *Let $B$ be a ruled cubic surface in $PG(4, q)$ with line directrix $t = \Sigma_\infty \cap B$. Then there is a unique spread $S$ of $\Sigma_\infty$ containing $t$ such that $B$ corresponds to a Baer subplane of $\mathcal{P}(S)$. Moreover, $S$ is necessarily a regular spread and thus $\mathcal{P}(S) \cong PG(2, q^2)$.*

*Proof.* Let $B$ be a ruled cubic surface with line directrix $t$, where $t = B \cap \Sigma_\infty$ and hence $t$ is a line of $\Sigma_\infty$. Let $C_1, C_2, \ldots, C_{q^2}$ be the conic directrices of $B$, and let $\pi_i$ be the plane containing $C_i$. Thus the intersections $\ell_i = \pi_i \cap \Sigma_\infty$, for $i = 1, 2, \ldots, q^2$, determine another $q^2$ lines of $\Sigma_\infty$. Recall that $t$ is skew to each plane $\pi_i$. Suppose that two planes, say $\pi_1$ and $\pi_2$, span a hyperplane $\Sigma$. Then the generators of $B$ joining points of $C_1$ to points of $C_2$ lie in $\Sigma$, and therefore $t$ lies in $\Sigma$. But this forces $t$ to meet $\pi_1$, a contradiction. Hence the planes $\pi_i$ and $\pi_j$ meet in a point, namely, the uniquely determined point $C_i \cap C_j$, for $i \neq j$. Since the only points of $\Sigma_\infty$ that lie on $B$ are the points of $t$, we see that $\ell_i$ and $\ell_j$ are necessarily skew lines for $i \neq j$. Therefore $t, \ell_1, \ell_2, \ldots, \ell_{q^2}$ form a set of $q^2 + 1$ mutually skew lines in $\Sigma_\infty$. That is, the Baer cubic $B$ induces a uniquely determined spread $S$ in $\Sigma_\infty$. Moreover, $B$ corresponds to a Baer subplane of $\mathcal{P}(S)$ in the Bruck-Bose representation (Theorem 3.20).

To show that $S$ is a regular spread, we need to show that it can be constructed from a pair of conjugate skew transversals in the quadratic extension $PG(3, q^2)$. Let $C$ be a conic directrix of $B$, and let $\ell = \pi \cap \Sigma_\infty$, where $\pi$ is the plane containing $C$. Then $\ell$ is an external line to $C$. Let $\overline{\ell}, \overline{t}$, and $\overline{C}$ denote the quadratic extensions of $\ell, t$, and $C$, respectively. Then $\overline{\ell}$ meets $\overline{C}$ in a pair of conjugate points $P, P^\sigma$ of $PG(4, q^2)$. Moreover, the quadratic extension $\overline{B}$ is a ruled cubic surface in $PG(4, q^2)$ with line directrix $\overline{t}$ that has $\overline{C}$ as one of its conic directrices.

Let $m$ and $m'$ be the generators of $\overline{B}$ through $P$ and $P^\sigma$, respectively. As they are generators of a ruled cubic surface, they are skew. Since the automorphic collineation $\sigma$ induced by the field automorphism $x \mapsto x^q$ fixes the (extended) generators of $B$ and creates a pairing of the generators of $\overline{B} \setminus B$, we see that the skew generators $m, m'$ form such a conjugate pair. This further implies that these generators are skew to $PG(4, q)$.

If $C_i$ is any conic directrix of $B$, then in the quadratic extension $PG(4, q^2)$, $\overline{C}_i$ is a conic directrix of $\overline{B}$ and hence meets the generators $m, m'$. Now the generators $m, m'$ meet $\overline{t}$ and $\overline{\ell}$, and thus they are contained in the quadratic extension $\overline{\Sigma}_\infty$ of $\Sigma_\infty$. For any conic directrix $C_i$, $i = 1, 2, \ldots, q^2$, the plane containing $\overline{C}_i$ meets $\overline{\Sigma}_\infty$ in the line $\overline{\ell}_i$, where $\ell_i$ is the previously defined line of the spread $S$. This line meets the conic $\overline{C}_i$ in two conjugate points in a similar fashion to the intersection of $\overline{C}$ and $\overline{\ell}$ above. As these are the only two points of $\overline{\Sigma}_\infty$ that lie on $\overline{C}_i$, we see that necessarily these two points are $m \cap \overline{\ell}_i$ and $m' \cap \overline{\ell}_i$. Thus each of the lines $m$ and $m'$ meets every (extended) line of $S$, and by Theorem 3.10 we know that $S$ is a regular spread of $\Sigma_\infty$ (uniquely determined by $m$ or $m'$). □

For a discussion of the Bruck-Bose representation of nonaffine Baer subplanes in non-Desarguesian planes, we refer the reader to the survey given by Vincenti [219], where a number of results from her article [218] are surveyed. Vincenti considers a certain class of non-Desarguesian translation planes and shows that certain varieties in $PG(4, q)$ correspond to nonaffine Baer subplanes.

Finally we discuss some results of Casse and Quinn [89] which enable us to determine whether a given ruled cubic surface of $PG(4, q)$ in the Bruck-Bose representation of $PG(2, q^2)$ is a Baer ruled cubic, and whether a given conic in $PG(4, q) \backslash \Sigma_\infty$ is a Baer conic. The answer is related to how the ruled cubic surface or conic meets the uniquely determined transversals of the associated regular spread $S$.

**Theorem 3.23.** *Let $B$ be a ruled cubic surface in $PG(4, q)$ with line directrix $t = B \cap \Sigma_\infty$ contained in the regular spread $S$ of $\Sigma_\infty$. Then $B$ corresponds to a Baer subplane of $P(S) \cong PG(2, q^2)$ tangent to $\ell_\infty$ if and only if the quadratic extension $\overline{B}$ of $B$ contains the transversals $g, g^\sigma$ of the regular spread $S$.*

*Proof.* By Theorem 3.22, it suffices to show that the number of ruled cubic surfaces in $PG(4, q)$ with line directrix a line of $S$ and containing $g, g^\sigma$ (the transversals of $S$) in the quadratic extension $PG(4, q^2)$ is equal to the number of nonaffine Baer subplanes of $PG(2, q^2)$. In the proof of Theorem 3.13, we showed that the number of affine Baer subplanes of $PG(2, q^2)$ is $q^3(q^2 + 1)(q + 1)$. In a similar way, we may count the total number of Baer subplanes in $PG(2, q^2)$ by using Theorem 2.8. Namely, the total number of Baer subplanes is equal to the number of unordered quadrangles in $PG(2, q^2)$ divided by the number of unordered quadrangles in $PG(2, q)$; that is,

$$\frac{(q^4 + q^2 + 1)(q^4 + q^2)q^4(q^4 - 2q^2 + 1)/4!}{(q^2 + q + 1)(q^2 + q)q^2(q^2 - 2q + 1)/4!} = (q^2 - q + 1)q^3(q^2 + 1)(q + 1).$$

Hence the number of nonaffine Baer subplanes in $PG(2, q^2)$ is

$$(q^2 - q + 1)(q^3)(q^2 + 1)(q + 1) - q^3(q^2 + 1)(q + 1) = q^4(q^4 - 1).$$

We now count ruled cubic surfaces with line directrix an element of $S$ whose quadratic extensions contain $g$ and $g^\sigma$. First pick an element $t$ of $S$, for which there are $q^2 + 1$ choices. Next choose a plane $\pi$ in $PG(4, q) \setminus \Sigma_\infty$ about a spread element distinct from $t$. There are $(q^2)(q^2) = q^4$ choices for $\pi$. Now choose an irreducible conic $\mathcal{C}$ in $\pi$ such that in the quadratic extension $PG(4, q^2)$, the extended conic $\overline{C}$ contains the two points $P = \overline{\pi} \cap g$ and $P^\sigma = \overline{\pi} \cap g^\sigma$. The line $\ell = PP^\sigma$ is fixed by $\sigma$ and hence meets $\pi$ in $q + 1$ points. Recalling that a conic is uniquely determined by any five of its points (necessarily having the property that no three are collinear), we see that any three noncollinear points of $\pi \setminus \ell$, together with $P$ and $P^\sigma$ will uniquely determine $\overline{C}$ and hence $\mathcal{C}$. The number of choices for $\mathcal{C}$ is

$$\frac{q^2(q^2 - 1)(q^2 - q)}{(q + 1)(q)(q - 1)} = q^3 - q^2.$$

Finally, we count the projectivities from $t$ to $\mathcal{C}$ that determine a ruled cubic surface whose quadratic extension will contain $g$ and $g^\sigma$. There are $q + 1$ choices for a line $m$ to be a line of $PG(4, q)$ joining a fixed point of $t$ to a point of $\mathcal{C}$. In the quadratic extension $PG(4, q^2)$, the three lines $g, g^\sigma$, and $\overline{m}$ determine a unique projectivity $\phi \in PGL(2, q^2)$ from the points of $\overline{t}$ to the points of $\overline{C}$. Furthermore, the set $\{g, g^\sigma, m\}$ is fixed by the automorphic collineation $\sigma$, and hence $\phi \in PGL(2, q)$. That is, $\phi$ determines a ruled cubic surface in $PG(4, q)$ of the desired type.

Note that a ruled cubic surface contains $q^2$ conic directrices, and any of these $q^2$ conics together with $t$ will yield the same ruled cubic surface for a given choice of $m$. Thus the total number of distinct Baer ruled cubic surfaces satisfying our conditions is precisely

$$(q^2 + 1)(q^4)(q^3 - q^2)(q + 1)/q^2 = q^4(q^4 - 1).$$

As this is equal to the number of nonaffine Baer subplanes of $PG(2, q^2)$ by our previous count, the result follows.  □

As an immediate corollary, we can determine which conics of $PG(4, q)$ are Baer conics; that is, we can determine which conics disjoint from $\Sigma_\infty$ correspond to Baer sublines of $PG(2, q^2)$ having no point on $\ell_\infty$.

**Corollary 3.24.** *Let $\mathcal{C}$ be an irreducible conic in a plane $\pi$ of $PG(4, q) \setminus \Sigma_\infty$ containing a line of the regular spread $\mathcal{S}$. Let $g, g^\sigma$ be the conjugate skew transversals of $\mathcal{S}$. Then $\mathcal{C}$ is a Baer conic in the Bruck-Bose representation of $\mathcal{P}(\mathcal{S}) \cong PG(2, q^2)$ if and only if in the quadratic extension $PG(4, q^2)$, $\overline{C}$ contains the two points $\overline{\pi} \cap g$ and $\overline{\pi} \cap g^\sigma$.*

### 3.4.3 Derivation in Bruck-Bose

Let $\mathcal{P}$ be a translation plane of dimension at most two over its kernel. Let $\Sigma_\infty$ be a hyperplane of $PG(4, q)$, and let $\mathcal{S}$ be the spread of $\Sigma_\infty$ that determines $\mathcal{P}$; that is, $\mathcal{P} = \mathcal{P}(\mathcal{S})$.

Suppose that $\mathcal{S}$ contains some regulus $\mathcal{R}$. Then we can construct a new spread by replacing the regulus $\mathcal{R}$ in $\mathcal{S}$ with its opposite regulus $\mathcal{R}'$, thereby obtaining the spread $\mathcal{S}' = (\mathcal{S} \setminus \mathcal{R}) \cup \mathcal{R}'$. We call this process **reversing a regulus**. Consider the plane $\mathcal{P}(\mathcal{S}')$ that corresponds to $\mathcal{S}'$. This plane is closely related to $\mathcal{P}(\mathcal{S})$. The regulus $\mathcal{R}$ corresponds to a Baer subline $b$ of $\mathcal{P}(\mathcal{S})$ which is contained in $\ell_\infty$. Planes of $\mathrm{PG}(4, q) \setminus \Sigma_\infty$ through a line of $\mathcal{R}'$ correspond to Baer subplanes of $\mathcal{P}(\mathcal{S})$ containing $b$. Hence $\mathcal{P}(\mathcal{S})$ and $\mathcal{P}(\mathcal{S}')$ have the same affine point set, while the affine lines of $\mathcal{P}(\mathcal{S}')$ are of two types. Lines of $\mathcal{P}(\mathcal{S})$ that meet $\ell_\infty$ in a point not in $b$ are also lines of $\mathcal{P}(\mathcal{S}')$. The other affine lines of $\mathcal{P}(\mathcal{S}')$ are Baer subplanes of $\mathcal{P}(\mathcal{S})$ containing $b$. Hence $\mathcal{P}(\mathcal{S}')$ is the plane obtained by deriving $\mathcal{P}(\mathcal{S})$ with respect to $b$ (see Section 3.2). That is, $\mathcal{P}(\mathcal{S}') \cong \mathcal{D}(\mathcal{P}(\mathcal{S}))$.

In particular, if $\mathcal{S}$ is regular, then reversing one regulus generates the Hall plane, while reversing "linear sets" of disjoint reguli generates the two-dimensional André planes. This process of reversing a regulus is discussed in [71, 73, 74].

### 3.4.4 Coordinates in Bruck-Bose

In this section we show how the coordinates of points in $\mathrm{PG}(2, q^2)$ relate to the coordinates of the corresponding points in the Bruck-Bose representation in $\mathrm{PG}(4, q)$. Let $\zeta$ be a primitive element in $\mathrm{GF}(q^2)$ with primitive polynomial

$$x^2 - t_1 x - t_0 \tag{3.1}$$

over $\mathrm{GF}(q)$. Then every element $\alpha \in \mathrm{GF}(q^2)$ can be uniquely written as $\alpha = a_0 + a_1 \zeta$ with $a_0, a_1 \in \mathrm{GF}(q)$. That is,

$$\mathrm{GF}(q^2) = \{x_0 + x_1 \zeta \mid x_0, x_1 \in \mathrm{GF}(q)\}.$$

Points in $\mathrm{PG}(2, q^2)$ have homogeneous coordinates

$$(X_0, X_1, X_2) = (x, y, z)$$

with $x, y, z \in \mathrm{GF}(q^2)$, not all zero. Let $\ell_\infty$ have equation $X_2 = z = 0$, so that affine points of $\mathrm{PG}(2, q^2)$ (that is, the points of $\mathrm{AG}(2, q^2)$) have coordinates $(x, y, 1)$, with $x, y \in \mathrm{GF}(q^2)$, using right normalized coordinates for $\mathrm{PG}(2, q^2)$. Lines in $\mathrm{PG}(2, q^2)$ have dual homogeneous coordinates $(\lambda, \mu, \nu)'$, where $\lambda, \mu, \nu \in \mathrm{GF}(q^2)$, not all zero.

Points in $\mathrm{PG}(4, q)$ have homogeneous coordinates

$$(X_0, X_1, X_2, X_3, X_4) = (x_0, x_1, y_0, y_1, z)$$

with $x_0, x_1, y_0, y_1, z \in \mathrm{GF}(q)$, not all zero. Let $\Sigma_\infty$ have equation $X_4 = z = 0$. Then the affine points of $\mathrm{PG}(4, q)$ (that is, the points of $\mathrm{PG}(4, q) \setminus \Sigma_\infty$) have coordinates $(x_0, x_1, y_0, y_1, 1)$, with $x_0, x_1, y_0, y_1 \in \mathrm{GF}(q)$, again using right normalized coordinates for $\mathrm{PG}(4, q)$. The hyperplanes of $\mathrm{PG}(4, q)$ have dual homogeneous coordinates $(a_0, a_1, a_2, a_3, a_4)'$, where $a_0, a_1, \ldots, a_4 \in \mathrm{GF}(q)$, not all zero.

Let $P$ be a point in $\mathrm{PG}(2,q^2)\backslash\ell_\infty$. Then we can write the coordinates of $P$ as

$$P = (\alpha,\beta,1) \qquad\qquad \text{for } \alpha,\beta \in \mathrm{GF}(q^2)$$
$$= (a_0 + a_1\zeta, b_0 + b_1\zeta, 1) \quad \text{for unique } a_0, a_1, b_0, b_1 \in \mathrm{GF}(q).$$

Consider the map

$$\varphi : \mathrm{PG}(2,q^2)\backslash\ell_\infty \longrightarrow \mathrm{PG}(4,q)\backslash\Sigma_\infty$$
$$\text{via} \quad (a_0 + a_1\zeta, b_0 + b_1\zeta, 1) \longmapsto (a_0, a_1, b_0, b_1, 1).$$

The map $\varphi$ is a bijection from the affine points of $\mathrm{PG}(2,q^2)$ to the affine points of $\mathrm{PG}(4,q)$ and will be called the **Bruck-Bose map**.

We now consider how $\varphi$ acts on the lines of $\mathrm{PG}(2,q^2)$. Since $\varphi$ is an affine map, we will work with affine equations of lines. Thus we group the lines of $\mathrm{PG}(2,q^2)$, distinct from $\ell_\infty$, into parallel classes of lines in $\mathrm{AG}(2,q^2)$.

Using left normalized coordinates, we represent the points on $\ell_\infty$ in $\mathrm{PG}(2,q^2)$ as

$$\ell_\infty = \{(1,\delta,0) \mid \delta \in \mathrm{GF}(q^2)\} \cup \{(0,1,0)\}.$$

The lines other than $\ell_\infty$ through $(0,1,0)$ have dual coordinates $(1,0,-\gamma)'$, for $\gamma \in \mathrm{GF}(q^2)$, while the lines other than $\ell_\infty$ through the point $(1,\delta,0)$ have dual coordinates $(-\delta,1,-\nu)'$, for $\nu \in \mathrm{GF}(q^2)$. These lines are shown in Figure 3.11.

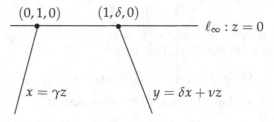

**Fig. 3.11.** Equations of lines in $\mathrm{PG}(2,q^2)$

We first compute the element of the (regular) spread $\mathcal{S}$ corresponding to the point $(0,1,0)$ on $\ell_\infty$ by applying the Bruck-Bose map to the affine lines of $\mathrm{PG}(2,q^2)$ through this point. Such a line, say $(1,0,-\gamma)'$, has affine equation $x = \gamma$, since we may assume $z = 1$ for affine points as above. We uniquely write $x = x_0 + x_1\zeta$, $\gamma = c_0 + c_1\zeta$ for $x_0, x_1, c_0, c_1 \in \mathrm{GF}(q)$. Thus we have the affine equation

$$x_0 + x_1\zeta = c_0 + c_1\zeta.$$

Equating like powers of $\zeta$ gives us $x_0 = c_0$ and $x_1 = c_1$. These are the affine equations of two hyperplanes in $\mathrm{PG}(4,q)$, whose corresponding projective equations are

$$x_0 - c_0 z = 0 \quad \text{and} \quad x_1 - c_1 z = 0.$$

These two equations define a plane (as two hyperplanes in $PG(4, q)$ intersect in a plane) which clearly is not in $\Sigma_\infty$. Hence the line $x = \gamma$ in $AG(2, q^2)$ corresponds to this plane in $PG(4, q)$. As $c_0, c_1$ vary over $GF(q)$ (or, equivalently, $\gamma$ varies over $GF(q^2)$), the resulting $q^2$ planes all contain the line defined as the intersection of the three hyperplanes whose equations are $x_0 = 0$, $x_1 = 0$, and $z = 0$. That is, this line in $\Sigma_\infty$ is the spread line that corresponds to the point $P_\infty = (0, 1, 0)$ on $\ell_\infty$ in $PG(2, q^2)$. Alternatively, this spread line could be written as

$$p_\infty = \langle (0, 0, 1, 0, 0), (0, 0, 0, 1, 0) \rangle,$$

using homogeneous coordinates for $PG(4, q)$.

We next compute the spread element corresponding to the point $(1, \delta, 0)$ of $\ell_\infty$ by taking a typical affine line in $PG(2, q^2)$ through this point, say with dual coordinates $(-\delta, 1, -v)'$, for some $v \in GF(q^2)$. To apply the Bruck-Bose map to this line, whose affine equation is $y = \delta x + v$, we uniquely write $\delta = d_0 + d_1 \zeta$, $v = n_0 + n_1 \zeta$, for $d_0, d_1, n_0, n_1 \in GF(q)$. Similarly, for the variables $x$ and $y$, we uniquely write $x = x_0 + x_1 \zeta$, $y = y_0 + y_1 \zeta$, where $x_0, x_1, y_0, y_1 \in GF(q)$. Substituting into the above affine equation, we obtain

$$
\begin{aligned}
y_0 + y_1 \zeta &= (d_0 + d_1 \zeta)(x_0 + x_1 \zeta) + (n_0 + n_1 \zeta) \\
&= d_0 x_0 + n_0 + (d_1 x_0 + d_0 x_1 + n_1)\zeta + (d_1 x_1)\zeta^2 \\
&= d_0 x_0 + n_0 + (d_1 x_0 + d_0 x_1 + n_1)\zeta + (d_1 x_1)(t_0 + t_1 \zeta)
\end{aligned}
$$

by using equation (3.1) for the primitive element $\zeta$ of $GF(q^2)$. Equating like powers of $\zeta$ yields two equations:

$$
\begin{aligned}
y_0 &= d_0 x_0 + t_0 d_1 x_1 + n_0 \\
y_1 &= d_1 x_0 + (d_0 + t_1 d_1)x_1 + n_1.
\end{aligned}
$$

These are the affine equations of two hyperplanes in $PG(4, q)$, whose corresponding homogeneous projective equations are

$$
\begin{aligned}
d_0 x_0 + t_0 d_1 x_1 - y_0 + n_0 z &= 0 \\
d_1 x_0 + (d_0 + t_1 d_1)x_1 - y_1 + n_1 z &= 0.
\end{aligned}
$$

These two equations define a plane which clearly is not in $\Sigma_\infty$. Hence the line $y = \delta x + v$ in $AG(2, q^2)$ corresponds to this plane in $PG(4, q)$. As $n_0, n_1$ vary over $GF(q)$ (or, equivalently, $v$ varies over $GF(q^2)$), the resulting $q^2$ planes all contain the line defined as the intersection of the three hyperplanes whose equations are

$$
\begin{aligned}
d_0 x_0 + t_0 d_1 x_1 - y_0 &= 0 \\
d_1 x_0 + (d_0 + t_1 d_1)x_1 - y_1 &= 0 \\
z &= 0.
\end{aligned}
$$

This line in $\Sigma_\infty$ is the spread line that corresponds to the point $(1, \delta, 0)$ on $\ell_\infty$ in $PG(2, q^2)$. Alternatively, this spread line could be written as

$$p_\delta = \langle (1,0,d_0,d_1,0), (0,1,t_0d_1,d_0+t_1d_1,0) \rangle,$$

using homogeneous coordinates for $PG(4,q)$.

Finally, allowing $\delta$ to vary over $GF(q^2)$, we obtain the spread

$$S = \{p_\infty\} \cup \{p_\delta \mid \delta \in GF(q^2)\}$$

of $\Sigma_\infty$, from which the translation plane $PG(2,q^2)$ can be obtained via the Bruck-Bose construction. In fact, since $PG(2,q^2)$ is Desarguesian, this spread must be regular.

This coordinate correspondence has been useful for studying curves in $PG(2,q^2)$ and their representations in $PG(4,q)$. Note that a similar correspondence holds for any translation plane of order $q^2$ coordinatized by a quasi-field with kernel $GF(q)$. For example, we can similarly relate the coordinates of points in the Hall plane $Hall(q^2)$ with points in its Bruck-Bose representation in $PG(4,q)$. This technique also generalizes to give coordinates for the Bruck-Bose representation of $PG(2,q^n)$ in $PG(2n,q)$ (see [36] for full details).

**Proof of Theorem 3.17**

We now consider a Baer subline of $PG(2,q^2)$ that is disjoint from $\ell_\infty$, and show that it corresponds to a conic in $PG(4,q)$.

*Proof.* Without loss of generality, we may consider our favorite Baer subline of the line $x = y$ in $PG(2,q^2)$. First note that the set

$$b' = \{(0,0,1) + d(1,1,1) \mid d \in GF(q) \cup \{\infty\}\}$$
$$= \{(d,d,d+1) \mid d \in GF(q)\} \cup \{(1,1,1)\}$$

is a Baer subline of the line $x = y$, and $b'$ meets $\ell_\infty$ in the point $(1,1,0)$. If we take the image of $b'$ under the homography given by

$$(x,y,z) \mapsto (x,y,(-\omega+1)x+\omega z)$$

for some $\omega \in GF(q^2) \setminus GF(q)$, we obtain the set

$$b = \{(d,d,d+\omega) \mid d \in GF(q)\} \cup \{(1,1,1)\}$$
$$= \{(d,d,d+w) \mid d \in GF(q) \cup \{\infty\}\}.$$

This is necessarily a Baer subline of the line $x = y$, and it is disjoint from $\ell_\infty$. Hence we use this Baer subline for the remainder of the proof.

Recall that $\zeta$ is a primitive element in $GF(q^2)$ with primitive polynomial $x^2 - t_1x - t_0$. The second root of this polynomial is $\zeta^q$, and therefore $\zeta + \zeta^q = t_1$ and $\zeta\zeta^q = -t_0$. Uniquely writing $\omega = w_1 + w_2\zeta$ for some $w_1, w_2 \in GF(q)$, with $w_2 \neq 0$, we see that $\omega^q = w_1 + w_2\zeta^q = w_1 + w_2t_1 - w_2\zeta$. Further recall that $\omega + \omega^q$ and $\omega\omega^q$ are both in $GF(q)$ from our discussion of norms and traces in Section 1.2. Hence for each $d \in GF(q)$, we have

$$(d + \omega)(d + \omega^q) = d^2 + ud + v \quad \text{for some } u, v \in \mathrm{GF}(q), \ v \neq 0.$$

Note that $d^2 + ud + v \neq 0$ for all choices of $d \in \mathrm{GF}(q)$.

Now consider the point $(d, d, d + \omega)$ of the Baer subline $b$, for some $d \in \mathrm{GF}(q) \cup \{\infty\}$. We have

$$(d, d, d + \omega) \equiv \left( d(d + \omega^q), d(d + \omega^q), (d + \omega)(d + \omega^q) \right)$$

$$\equiv \left( \frac{d^2 + dw_1 + dt_1 w_2 - dw_2 \zeta}{d^2 + ud + v}, \frac{d^2 + dw_1 + dt_1 w_2 - dw_2 \zeta}{d^2 + ud + v}, 1 \right),$$

where $\equiv$ is used to denote homogeneous coordinates that represent the same point. In fact, this point of $\mathrm{PG}(2, q^2)$ corresponds to the following point of $\mathrm{PG}(4, q)$, which lies in the plane with equations $x_0 = y_0$ and $x_1 = y_1$:

$$\left( \frac{d^2 + dw_1 + dt_1 w_2}{d^2 + ud + v}, \frac{-dw_2}{d^2 + ud + v}, \frac{d^2 + dw_1 + dt_1 w_2}{d^2 + ud + v}, \frac{-dw_2}{d^2 + ud + v}, 1 \right).$$

Hence the $q + 1$ points of $b$ correspond to a set of $q + 1$ points in $\mathrm{PG}(2, q)$ projectively equivalent to

$$\mathcal{C} = \left\{ \left( d^2 + dw_1 + dt_1 w_2, \ -dw_2, \ d^2 + ud + v \right) \mid d \in \mathrm{GF}(q) \cup \{\infty\} \right\}.$$

Since $\left( d^2 + dw_1 + dt_1 w_2, \ -dw_2, \ d^2 + ud + v \right)$ is the image of $(d^2, d, 1)$ under the homography

$$(x, y, z) \longmapsto (x + (w_1 + t_1 w_2)y, \ -w_2 y, \ x + uy + vz),$$

we see that the set $\mathcal{C}$ is projectively equivalent to the set

$$\{ (d^2, d, 1) \mid d \in \mathrm{GF}(q) \cup \{\infty\} \},$$

and hence $\mathcal{C}$ is an irreducible conic in $\mathrm{PG}(2, q)$. $\qquad\square$

# 4

# Unitals Embedded in Desarguesian Planes

In this chapter we discuss Buekenhout's two constructions for unitals embedded in two-dimensional translation planes, and then apply these constructions in the Desarguesian plane of order $q^2$. One of these constructions produces only the classical unital (that is, the Hermitian curve), but the other produces several inequivalent unitals. These unitals are enumerated and their stabilizer subgroups are determined. We also discuss various geometric properties of these unitals.

## 4.1 Buekenhout Constructions

Buekenhout [81] used the Bruck-Bose representation, as introduced in Section 3.4, to investigate unitals. His investigation led to constructions of non-classical unitals in $PG(2, q^2)$ and certain unitals in non-Desarguesian planes. Furthermore, many authors have built upon his work to develop various characterizations of unitals (see Chapter 7).

We use the notation for the Bruck-Bose representation given in Section 3.4. If $U$ is a unital in a translation plane $\mathcal{P}$ of dimension at most two over its kernel, then we let $\mathcal{U}$ denote the corresponding set of points in the Bruck-Bose representation of $\mathcal{P}$ in $PG(4, q)$. Buekenhout [81] showed that the set $\mathcal{U}$ corresponding to a classical unital $U$ of $PG(2, q^2)$ is a quadric in $PG(4, q)$. The cases when $U$ is tangent to $\ell_\infty$ and when $U$ is secant to $\ell_\infty$ are treated separately.

**Theorem 4.1.** *Let $U$ be a classical unital in $\mathcal{P}(\mathcal{S}) \cong PG(2, q^2)$.*

1. *If $U$ meets $\ell_\infty$ in a point $P_\infty$, then $\mathcal{U}$ is an elliptic cone whose vertex lies on the corresponding line $p_\infty$ of $\mathcal{S}$.*
2. *If $U$ meets $\ell_\infty$ in a Baer subline $b$, then $\mathcal{U}$ is a nonsingular quadric that meets $\mathcal{S}$ in the regulus corresponding to $b$.*

S. Barwick, G. Ebert, *Unitals in Projective Planes*,
DOI: 10.1007/978-0-387-76366-8_4, © Springer Science+Business Media, LLC 2008

*Proof.* We use the notation for coordinates in the Bruck-Bose representation that is given in Section 3.4.4. That is, we let $\zeta$ be a primitive element in $GF(q^2)$ with primitive polynomial

$$x^2 - t_1 x - t_0.$$

Recall that a point $(x, y, 1) = (x_0 + x_1\zeta, y_0 + y_1\zeta, 1)$ in $AG(2, q^2)$ corresponds to the point $(x_0, x_1, y_0, y_1, 1)$ in $PG(4, q)\setminus \Sigma_\infty$.

Let $U$ be a classical unital in $PG(2, q^2)$ which is tangent to $\ell_\infty$. Then, without loss of generality, we may assume that $U$ has affine equation

$$xx^q + y + y^q = 0,$$

with $U \cap \ell_\infty = (0, 1, 0)$. We convert this to an equation in $PG(4, q)$ by substituting $x = x_0 + x_1\zeta$, $y = y_0 + y_1\zeta$. Note that $x_i, y_i \in GF(q)$, so that $x_i^q = x_i$ and $y_i^q = y_i$. Furthermore, as $\zeta$ is a root of $x^2 - t_1 x - t_0$, so is $\zeta^q$. Hence $x^2 - t_1 x - t_0 = (x - \zeta)(x - \zeta^q)$ and therefore $\zeta^q + \zeta = t_1$, $\zeta^q\zeta = -t_0$. Thus from the equation for $U$ we have

$$\begin{aligned} 0 &= (x_0 + x_1\zeta)(x_0 + x_1\zeta)^q + (y_0 + y_1\zeta) + (y_0 + y_1\zeta)^q \\ &= (x_0 + x_1\zeta)(x_0 + x_1\zeta^q) + (y_0 + y_1\zeta) + (y_0 + y_1\zeta^q) \\ &= x_0^2 + t_1 x_0 x_1 - t_0 x_1^2 + 2y_0 + t_1 y_1. \end{aligned}$$

This is the affine equation of a quadric $\mathcal{U}$ in $PG(4, q)$, whose projective homogeneous equation is

$$x_0^2 + t_1 x_0 x_1 - t_0 x_1^2 + 2y_0 z + t_1 y_1 z = 0.$$

Thus the affine points of $U$ correspond precisely to the affine points of $\mathcal{U}$. The points in $\Sigma_\infty$ on the quadric $\mathcal{U}$ satisfy the equation $x_0^2 + t_1 x_0 x_1 - t_0 x_1^2 = 0$. Since $x^2 - t_1 x - t_0$ is irreducible over $GF(q)$, these points constitute the line $x_0 = x_1 = z = 0$; namely, $\Sigma_\infty \cap \mathcal{U}$ is the spread line $p_\infty$ of $S$ corresponding to the point $P_\infty = (0, 1, 0)$ of $\ell_\infty$. Hence the points of this classical unital $U$ correspond precisely to the points of the quadric $\mathcal{U}$ in $PG(4, q)$.

Note that if $t_1 = 0$, then $\zeta^2 = t_0 \in GF(q)$ and so $\zeta^{2(q-1)} = 1$, hence the order of $\zeta$ divides $2(q - 1)$. This contradicts our choice of $x^2 - t_1 x - t_0$ as a primitive polynomial for $GF(q^2)$. Hence we have $t_1 \neq 0$.

Intersecting $\mathcal{U}$ with the hyperplane whose equation is $y_0 = 0$, one obtains a three-dimensional elliptic quadric as described in Section 1.3. The point $V = (0, 0, 1, 0, 0)$ on $\mathcal{U}$ does not lie on this elliptic quadric. In fact, straightforward computations show that $\mathcal{U}$ is an elliptic cone whose vertex is this point $V$ and whose base may be taken as the above elliptic quadric. As $V$ lies on the spread line $p_\infty$, the proof of (1) is complete.

Now consider the case when the classical unital $U$ is secant to $\ell_\infty$. Then, without loss of generality, we may assume that $U$ has affine equation

$$xx^q + yy^q + 1 = 0,$$

with points at infinity $(x, y, 0)$ satisfying $xx^q + yy^q = 0$. That is,

$$U \cap \ell_\infty = \{P_\delta = (1, \delta, 0) \mid \delta^{q+1} = -1\}.$$

Computations similar to those made above show that in $PG(4, q)$ this corresponds to a quadric $U$ whose projective homogeneous equation is

$$x_0^2 + t_1 x_0 x_1 - t_0 x_1^2 + y_0^2 + t_1 y_0 y_1 - t_0 y_1^2 + z^2 = 0.$$

This quadric is nonsingular and intersects $\Sigma_\infty$ in the three-dimensional hyperbolic quadric of equation $x_0^2 + t_1 x_0 x_1 - t_0 x_1^2 + y_0^2 + t_1 y_0 y_1 - t_0 y_1^2 = 0$ (for instance, see Theorem 22.2.1 in [134]). Straightforward, but messy, computations show that this hyperbolic quadric contains the $q + 1$ spread lines $\{p_\delta \mid \delta^{q+1} = -1\}$ of the regular spread $S$. These spread lines necessarily form a regulus as they lie on a hyperbolic quadric. Moreover, they correspond precisely to the points of $U \cap \ell_\infty$. This completes the proof of (2).    □

Buekenhout [81] also investigated the converse of this result, which led to the first examples of nonclassical unitals in $PG(2, q^2)$.

**Theorem 4.2.** *Let $S$ be a spread, not necessarily regular, in a hyperplane $\Sigma_\infty$ of $PG(4, q)$.*

1. *If $U$ is an ovoidal cone of $PG(4, q)$ that meets $\Sigma_\infty$ in a line of $S$, then $U$ corresponds to a unital in $P(S)$ which is tangent to $\ell_\infty$.*
2. *If $U$ is a nonsingular quadric in $PG(4, q)$ that meets $\Sigma_\infty$ in a regulus of the spread $S$, then $U$ corresponds to a unital in $P(S)$ which is secant to $\ell_\infty$.*

*Proof.* Suppose $U$ is an ovoidal cone that meets $\Sigma_\infty$ in a line $p_\infty$ of $S$, so that $U$ necessarily has its vertex $V$ on $p_\infty$. As an ovoid contains $q^2 + 1$ points, an ovoidal cone contains $(q^2 + 1)q + 1 = q^3 + q + 1$ points. Since $U$ meets $\Sigma_\infty$ in a spread line, $U$ corresponds to a set $U$ with $q^3 + q + 1 - (q + 1) + 1 = q^3 + 1$ points in $P(S)$. Let $P_\infty$ be the point of $\ell_\infty$ corresponding to the spread line $p_\infty$ and let $\ell$ be a line of $P(S)$ through $P_\infty$ distinct from $\ell_\infty$. Then $\ell$ corresponds to a plane $\pi$ of $PG(4, q) \setminus \Sigma_\infty$ containing $p_\infty$.

Since $\Sigma_\infty$ is the unique tangent hyperplane to $U$ at $p_\infty$ and $\pi$ is not contained in $\Sigma_\infty$, by Theorem 1.19 we know that $\pi$ is not a tangent plane at $p_\infty$. Thus $\pi$ meets $U$ in $p_\infty$ and another line $m$ through $V$. Hence in $P(S)$, $\ell$ meets $U$ in $q + 1$ points. Now let $\ell$ be a line of $P(S)$ that meets $\ell_\infty$ in a point distinct from $P_\infty$. Then $\ell$ corresponds to a plane $\pi$ of $PG(4, q) \setminus \Sigma_\infty$ containing a spread line distinct from $p_\infty$. This plane contains no lines of $U$, and thus meets $U$ in either a point of $PG(4, q) \setminus \Sigma_\infty$ or a conic of $PG(4, q) \setminus \Sigma_\infty$. Hence $\ell$ meets $U$ in 1 or $q + 1$ points. Finally, $\ell_\infty$ meets $U$ in one point, $P_\infty$. Therefore by Definition 2.13, $U$ is a unital of $P(S)$ and (1) is proved.

Next suppose that $U$ is a nonsingular quadric in $PG(4, q)$ which meets $\Sigma_\infty$ in a regulus of the spread $S$. As discussed in Section 1.3, $U$ contains $(q^2 + 1)(q + 1)$ points with $(q + 1)^2$ of them lying on the regulus of $S$

which covers the points of $\mathcal{U} \cap \Sigma_\infty$. Hence $\mathcal{U}$ corresponds to a set $U$ with $(q^2 + 1)(q + 1) - (q + 1)^2 + (q + 1) = q^3 + 1$ points in $\mathcal{P}(\mathcal{S})$. Let $\ell$ be a line of $\mathcal{P}(\mathcal{S})$ that meets $\ell_\infty$ in a point of $U \cap \ell_\infty$. Then $\ell$ corresponds to a plane $\pi$ of $\mathrm{PG}(4, q) \backslash \Sigma_\infty$ that contains a line of $\mathcal{U} \cap \Sigma_\infty$. As every plane meets $\mathcal{U}$ in a point, a line, a pair of intersecting lines, or an irreducible conic (the intersection must be a planar quadric, possibly degenerate), necessarily $\pi$ meets $\mathcal{U}$ in one or two lines, one of which is a spread line. Hence $\ell$ meets $U$ in one point or $q + 1$ points. If $\ell$ is a line of $\mathcal{P}(\mathcal{S})$ that meets $\ell_\infty$ in a point not in $U$, then $\ell$ corresponds to a plane $\pi$ of $\mathrm{PG}(4, q) \backslash \Sigma_\infty$ which meets $\Sigma_\infty$ in a spread line which is disjoint from $\mathcal{U}$. Hence $\pi$ meets $\mathcal{U}$ in a point or a conic contained in $\mathrm{PG}(4, q) \backslash \Sigma_\infty$. Thus $\ell$ meets $U$ in 1 or $q + 1$ points. Finally, $\ell_\infty$ meets $U$ in $q + 1$ points. Therefore $U$ is a unital of $\mathcal{P}(\mathcal{S})$ by definition, completing the proof.    $\square$

For a unital $U$ in $\mathcal{P}(\mathcal{S})$ that is constructed from an ovoidal cone, Corollary 3.14 implies that the $q^2$ secant lines of $U$ through $U \cap \ell_\infty$ meet $U$ in Baer sublines. Of course, for classical unitals of $\mathrm{PG}(2, q^2)$ every secant line intersection is a Baer subline. The above theorems also imply the existence of nonclassical unitals in $\mathrm{PG}(2, q^2)$, for certain values of $q$. Namely, let $q \geq 8$ be an odd power of 2. In the Bruck-Bose representation of $\mathrm{PG}(2, q^2)$, let $\mathcal{U}$ be an ovoidal cone with vertex $V$ in $\Sigma_\infty$ and base a Suzuki-Tits ovoid that meets $\Sigma_\infty$ in a point on the spread line containing $V$. Then $\mathcal{U}$ corresponds to a nonclassical unital in $\mathrm{PG}(2, q^2)$ by Theorems 4.1 and 4.2. We call such a unital a **Buekenhout-Tits unital**.

Note that Buekenhout's ovoidal cone construction gives rise to unitals in every translation plane of dimension at most two over its kernel. Furthermore, Buekenhout's nonsingular quadric construction gives rise to a unital in every such translation plane $\mathcal{P}(\mathcal{S})$ in which the spread $\mathcal{S}$ contains a regulus.

Metz [172] showed that nonclassical unitals exist in $\mathrm{PG}(2, q^2)$ for all $q > 2$. His proof proceeds by first using a nice counting argument to show that there exists a non-Baer conic in $\mathrm{PG}(4, q)$ (see Theorem 3.18), and then showing that there is an ovoid and hence an ovoidal cone containing this conic. The resulting unital obtained from Buekenhout's ovoidal cone construction turns out to be nonclassical. We now provide the details of this argument. It should be noted that in 2000, Brown [68] proved that any ovoid containing a conic is an elliptic quadric. Hence these nonclassical unitals of Metz necessarily arise from elliptic (orthogonal) cones.

**Lemma 4.3.** *Let $C$ be an irreducible conic in a plane $\pi$ of $\mathrm{PG}(3, q)$ and let $\ell$ be a line of $\pi$ external to $C$. Let $\pi'$ be another plane through $\ell$. If $P$ is a point in $\pi' \backslash \ell$, then there is an ovoid $\mathcal{O}$ of $\mathrm{PG}(3, q)$ that contains $C$ and $P$ and has $\pi'$ as a tangent plane.*

*Proof.* Let $\mathcal{O}'$ be any ovoid containing $C$ (as noted above, by Brown's recent result this ovoid must be an elliptic quadric). As $\ell$ is external to $C$, $\ell$ is also external to $\mathcal{O}'$. A simple counting argument now shows that the $q + 1$ planes through $\ell$ consist of two tangent planes to $\mathcal{O}'$ and $q-1$ planes meeting $\mathcal{O}'$ in an oval, one of these being $\pi$. In particular, $\ell$ is contained in some plane $\tau$ which is tangent to $\mathcal{O}'$, say at the point $Q$. Now, if the point $P$ and plane $\pi'$ are given as in the statement of the lemma, then there is an elation of $PG(3, q)$ with axis $\pi$ and center $PQ \cap \pi$ which maps $Q$ to $P$, $\tau$ to $\pi'$, and $\mathcal{O}'$ to an ovoid $\mathcal{O}$ through $\mathcal{C}$ which is tangent to $\pi'$ at $P$, thus proving the result.        □

**Theorem 4.4.** *If $q$ is a prime power, $q > 2$, then there exists a nonclassical unital in* $PG(2, q^2)$.

*Proof.* Let $\mathcal{S}$ be a regular spread in a hyperplane $\Sigma_\infty$ of $PG(4, q)$, so that we have $\mathcal{P}(\mathcal{S}) \cong PG(2, q^2)$. Let $\pi$ be a plane of $PG(4, q) \backslash \Sigma_\infty$ containing a line $\ell$ of $\mathcal{S}$. Then $\pi$ corresponds to a line of $\mathcal{P}(\mathcal{S})$, and by Theorem 3.18 there is a conic $\mathcal{C}$ of $\pi \backslash \ell$ that does not correspond to a Baer subline of $\mathcal{P}(\mathcal{S})$. Let $V$ be a point of $\Sigma_\infty \backslash \ell$ as shown in Figure 4.1.

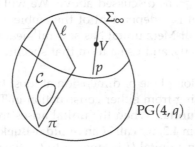

**Fig. 4.1.** Proof of Theorem 4.4

Consider the incidence structure $\Pi_V$ defined as follows: *points* are the lines of $PG(4, q)$ through $V$; *lines* are the planes of $PG(4, q)$ through $V$; *planes* are the 3-spaces of $PG(4, q)$ through $V$; and incidence is inherited from $PG(4, q)$. Then $\Pi_V$ is isomorphic to $PG(3, q)$; in fact, this geometry is usually called the **quotient space** of $V$. Note that to prove $\Pi_V$ is a 3-space, one can easily verify the following four axioms: two distinct points of $\Pi_V$ lie on a unique line of $\Pi_V$; two distinct planes of $\Pi_V$ meet in a unique line of $\Pi_V$; three noncollinear points of $\Pi_V$ lie in a unique plane of $\Pi_V$; three planes of $\Pi_V$ not through a line meet in a unique point of $\Pi_V$.

In $\Pi_V$ the lines through $V$ that meet $\pi$ form a plane $\pi_V$, and the lines of $\Sigma_\infty$ through $V$ form another plane $\pi_\infty$. Furthermore, the $q+1$ lines through $V$ that meet the spread line $\ell$ form the line $\ell_V = \pi_\infty \cap \pi_V$. Finally, consider the spread line $p$ through $V$. It corresponds to a point $P_V$ of $\Pi_V$ in the plane $\pi_\infty$, and $P_V \notin \ell_V$. Note that the $q+1$ lines through $V$ that meet $\mathcal{C}$ form a conic $\mathcal{C}_V$ in $\pi_V$. Hence the hypotheses of Lemma 4.3 are satisfied. Namely, $\mathcal{C}_V$ is a conic in the plane $\pi_V$, $\ell_V$ is a line in $\pi_V$ which is external to $\mathcal{C}_V$, $\pi_\infty$ is another plane through $\ell_V$, and $P_V$ is a point in $\pi_\infty \backslash \ell_V$. Thus $\Pi_V$ contains an ovoid $\mathcal{O}_V$ through $\mathcal{C}_V$ and $P_V$ which is tangent to the plane $\pi_\infty$. In $PG(4, q)$ this corresponds to an ovoidal cone with vertex $V$ which meets $\Sigma_\infty$ in the spread line $p$ of $\mathcal{S}$. Thus, by Theorem 4.2 the ovoidal cone corresponds to a unital $U$ in $\mathcal{P}(\mathcal{S}) \cong PG(2, q^2)$. Moreover, since the base of this ovoidal cone is an ovoid containing a non-Baer conic, the line of $PG(2, q^2)$ corresponding to $\pi$ meets $U$ in $q + 1$ points that do not form a Baer subline. Recall from our discussion in Section 2.2 that every secant line of the classical unital is a Baer subline. Thus $U$ is not a classical unital.                                   $\square$

One finds in the literature a wide variety of names for the different types of unitals arising from Buekenhout's two constructions. Because of the above theorem of Metz, unitals in $PG(2, q^2)$ arising from Buekenhout's ovoidal cone construction are often called "Buekenhout-Metz" unitals. However, some authors have used the term Buekenhout-Metz to refer only to the unitals arising from an elliptic cone. Moreover, the terminology Buekenhout-Metz is often used when the unitals are constructed in non-Desarguesian translation planes of order $q^2$, as discussed above. We will try to use consistent nomenclature, one that is independent of the ambient plane for the unital. As the term Buekenhout-Metz unital has several meanings in the literature, we avoid using it directly, and use a term that specifies the type of cone involved.

Let $\mathcal{P}$ be a translation plane of dimension at most two over its kernel. If $U$ is a unital in $\mathcal{P}$ arising from either construction in Theorem 4.2, then we call $U$ a **Buekenhout unital**. When the unital arises from the ovoidal cone construction in Theorem 4.2, we call $U$ an **ovoidal-Buekenhout-Metz unital**. To emphasize that such a unital $U$ is tangent to $\ell_\infty$, we sometimes refer to $U$ as an *ovoidal-Buekenhout-Metz unital with respect to the line $\ell_\infty$*. If $P_\infty = U \cap \ell_\infty$, since there is a unique tangent line through any point of $U$, we also call $U$ an *ovoidal-Buekenhout-Metz unital with respect to the point $P_\infty$*. If the ovoidal cone is an elliptic (orthogonal) cone, we call the unital an **orthogonal-Buekenhout-Metz unital**. As defined above, a unital arising from an ovoidal cone with base a Tits ovoid is called a **Buekenhout-Tits unital**. On the other hand, a unital arising from the nonsingular quadric construction of Buekenhout will be called a **nonsingular-Buekenhout unital**. For a complete discussion of names currently used in the literature, see Appendix A.

We now look at nonsingular-Buekenhout unitals; that is, unitals that correspond to nonsingular quadrics of $PG(4, q)$ which meet $\Sigma_\infty$ in a regulus of $\mathcal{S}$. Barwick [33] used a counting argument to show that in the Desarguesian plane, any unital constructed in this way is necessarily classical. In particular, Buekenhout's construction using nonsingular quadrics of $PG(4, q)$ does not give new examples of unitals in $PG(2, q^2)$.

**Theorem 4.5.** *If $U$ is a nonsingular-Buekenhout unital in $PG(2, q^2)$, then $U$ is classical.*

*Proof.* First we count the number of classical unitals in $PG(2, q^2)$ that contain a given Baer subline of $\ell_\infty$. We do this by counting in two ways the

ordered pairs $(\ell, U)$, where $U$ is a classical unital of $PG(2, q^2)$ and $\ell$ is a secant line of $U$. On the one hand, the number of classical unitals in $PG(2, q^2)$ is
$q^3(q^2 + 1)(q^3 - 1)$ (see [96], for instance), and each such unital has $q^4 - q^3 + q^2$
secant lines by Theorem 2.3. On the other hand, there are $q^4 + q^2 + 1$ lines
in $PG(2, q^2)$, and each of these lines is secant to the same number of classical unitals since $PGL(3, q^2)$ acts transitively on lines in $PG(2, q^2)$. Hence the
number of classical unitals secant to a given line of $PG(2, q^2)$, such as $\ell_\infty$, is

$$\frac{q^3(q^2 + 1)(q^3 - 1)(q^4 - q^3 + q^2)}{q^4 + q^2 + 1} = q^5(q^2 + 1)(q - 1).$$

By Theorem 2.6 any three distinct points of $PG(1, q^2)$ determine a unique
Baer subline. Hence $\binom{q^2+1}{3} / \binom{q+1}{3} = q(q^2 + 1)$ is the number of Baer sublines
on a line of $PG(2, q^2)$. Using the triple transitivity of $PGL(2, q^2)$ on the points
of $PG(1, q^2)$, we see that the number of classical unitals containing a given
Baer subline of $\ell_\infty$ is

$$\frac{q^5(q^2 + 1)(q - 1)}{q(q^2 + 1)} = q^4(q - 1).$$

We now count the number of nonsingular quadrics in $PG(4, q)$ that pass
through a prescribed regulus of $\Sigma_\infty$ by first counting in two ways the ordered
pairs $(\mathcal{Q}, \Sigma)$, where $\Sigma$ is a hyperplane of $PG(4, q)$ and $\mathcal{Q}$ is a nonsingular
quadric in $PG(4, q)$ that meets $\Sigma$ in a hyperbolic quadric. On the one hand,
there are $q^6(q^3 - 1)(q^5 - 1)$ nonsingular quadrics in $PG(4, q)$, and each such
quadric has $(q^4 + q^2)/2$ hyperplane sections which are hyperbolic quadrics
(see Theorems 22.6.2 and 22.8.2 in [134]). On the other hand, there are $q^4 +
q^3 + q^2 + q + 1$ hyperplanes in $PG(4, q)$, and each such hyperplane meets the
same number of nonsingular quadrics of $PG(4, q)$ in a hyperbolic quadric
since the automorphism group of $PG(4, q)$ acts transitively on hyperplanes.
Thus the number of nonsingular quadrics of $PG(4, q)$ meeting any particular
hyperplane, such as $\Sigma_\infty$, in a hyperbolic quadric is

$$\frac{\frac{1}{2}(q^4 + q^2)(q^6)(q^3 - 1)(q^5 - 1)}{q^4 + q^3 + q^2 + q + 1} = \frac{1}{2}q^8(q^2 + 1)(q - 1)(q^3 - 1).$$

Since the number of hyperbolic quadrics in $PG(3, q)$ is $q^4(q^2 + 1)(q^3 - 1)/2$
(see Theorem 22.6.2 in [134]), another transitivity argument shows that the
number of nonsingular quadrics meeting $\Sigma_\infty$ in a given hyperbolic quadric
(hence in a given regulus) is

$$\frac{\frac{1}{2}q^8(q^2 + 1)(q - 1)(q^3 - 1)}{\frac{1}{2}q^4(q^2 + 1)(q^3 - 1)} = q^4(q - 1).$$

As shown above, this is the same number as the number of classical unitals
containing a given Baer subline of $\ell_\infty$.

Thus the number of nonsingular-Buekenhout unitals in $PG(2, q^2)$ that contain a given Baer subline of $\ell_\infty$ is the same as the number of classical unitals passing through that Baer subline. Hence in $PG(2, q^2)$ all nonsingular-Buekenhout unitals are classical.                                                    □

In Section 2.3, we showed that the dual of a unital is a unital in the dual plane. As the Desarguesian plane is self-dual, the dual of a unital in $PG(2, q^2)$ is a unital in $PG(2, q^2)$. Several authors have contributed to showing that the dual of a Buekenhout unital in $PG(2, q^2)$ is a Buekenhout unital ([22, 87, 108, 192]), so the dual construction does not give any new unitals in $PG(2, q^2)$.

To date there are relatively few known constructions of unitals in the Desarguesian plane $PG(2, q^2)$; namely, the constructions described in this section and various special constructions of the classical unital. Hence the only known unitals in the Desarguesian plane are the ovoidal-Buekenhout-Metz unitals, one of which is the classical unital. Penttila and Royle [178] showed that the only unitals in $PG(2, 9)$ are ovoidal-Buekenhout-Metz unitals. An interesting open question is the following: Are there any other unitals in $PG(2, q^2)$?

## 4.2 Unitals Embedded in $PG(2, q^2)$

In this section we carefully describe the unitals embedded in the Desarguesian projective plane $PG(2, q^2)$ which are obtained from one of Buekenhout's two methods. As discussed above, only the classical unital of $PG(2, q^2)$ can be obtained if one starts with a nonsingular quadric of $PG(4, q)$. Thus we concentrate on the ovoidal-Buekenhout-Metz unitals embedded in $PG(2, q^2)$.

### 4.2.1 The Odd Characteristic Case

In this section we follow the approach of Baker and Ebert [22]. Assume that $q$ is an odd prime power, say $q = p^e$ for some odd prime $p$, and consider the class of ovoidal-Buekenhout-Metz unitals embedded in $PG(2, q^2)$. For odd $q$, as discussed in Section 1.4, every ovoidal cone in $PG(4, q)$ is an elliptic cone (see [28] or [176]) and hence any ovoidal-Buekenhout-Metz unital in $PG(2, q^2)$ is necessarily an orthogonal-Buekenhout-Metz unital. Moreover, as discussed in Section 1.3, any two elliptic cones are projectively equivalent.

We use the coordinates for the Bruck-Bose representation of $PG(2, q^2)$, as discussed in Section 3.4.4, but with a different choice for the basis of $GF(q^2)$ over the subfield $GF(q)$. Namely, instead of using the basis $\{1, \zeta\}$, where $\zeta$ is a primitive element of $GF(q^2)$, we use the basis $\{1, \epsilon\}$, where $\epsilon = \zeta^{(q+1)/2}$. Hence $\epsilon^2 = w$ is a primitive element of the subfield $GF(q)$, and $\epsilon^q = -\epsilon$. This simply makes some of the computations easier for odd $q$. Using Greek letters for elements of $GF(q^2)$ and Latin letters for elements of $GF(q)$, we write $\alpha = a_0 + a_1\epsilon$, $\beta = b_0 + b_1\epsilon$, $\delta = d_0 + d_1\epsilon$, and so on, where $\alpha, \beta, \delta \in GF(q^2)$.

Thus the (affine) point $(\alpha, \beta, 1)$ of PG$(2, q^2) \backslash \ell_\infty$ corresponds to the (affine) point $(a_0, a_1, b_0, b_1, 1)$ of PG$(4, q) \backslash \Sigma_\infty$. The point (at infinity) $P_\infty = (0, 1, 0)$ on $\ell_\infty$ corresponds to the line $p_\infty = \langle (0, 0, 1, 0, 0), (0, 0, 0, 1, 0) \rangle$ in the regular spread $S$ of $\Sigma_\infty$, while the point $P_\delta = (1, \delta, 0)$ on $\ell_\infty$ corresponds to the line $p_\delta = \langle (1, 0, d_0, d_1, 0), (0, 1, wd_1, d_0, 0) \rangle$ in this regular spread. It should be observed that in the notation of Section 3.4.4 we have $t_0 = w$ and $t_1 = 0$ since $\epsilon^2 - w = 0$.

As in Section 1.3, every elliptic cone in PG$(4, q)$ is projectively equivalent to one with an equation of the form $f(X_0, X_1) + X_3 X_4 = 0$, where $f$ is an irreducible binary quadratic form. The vertex of this cone is $V = (0, 0, 1, 0, 0)$, and the unique tangent hyperplane (see Theorem 1.17) at the generator $p_\infty = \langle (0, 0, 1, 0, 0), (0, 0, 0, 1, 0) \rangle$ of this cone is $\Sigma_\infty$, whose equation is $X_4 = 0$. Since $p_\infty$ is a line of the above regular spread $S$, this elliptic cone will correspond to an orthogonal-Buekenhout-Metz unital of PG$(2, q^2)$ with $\ell_\infty$ tangent to the unital at the point $P_\infty = (0, 1, 0)$. In fact, every orthogonal-Buekenhout-Metz unital embedded in PG$(2, q^2)$ will be projectively equivalent to one of this form, for some irreducible binary quadratic form $f(X_0, X_1)$. Namely, once the regular spread $S$ of $\Sigma_\infty$ is chosen, and hence PG$(2, q^2)$ is constructed, we may use the transitivity of the associated groups to choose without loss of generality which (unique) spread line will be the intersection of the elliptic cone with $\Sigma_\infty$ and which point on this spread line will be the vertex of the cone. Choosing the spread line $p_\infty$ and the vertex $V$ above, the elliptic cone may be represented in the above form.

In order to sort out equivalences as the irreducible binary quadratic form $f(X_0, X_1)$ varies, we introduce coordinates for the resulting orthogonal-Buekenhout-Metz unitals. To that end, let $\alpha, \beta \in$ GF$(q^2)$ and define

$$U_{\alpha\beta} = \{(x, \alpha x^2 + \beta x^{q+1} + r, 1) \mid x \in \mathrm{GF}(q^2), r \in \mathrm{GF}(q)\} \cup \{(0, 1, 0)\}. \quad (4.1)$$

Using the basis $\{1, \epsilon\}$ for GF$(q^2)$ over GF$(q)$ as above, we see that the affine point $(x, y, 1)$ of PG$(2, q^2) \backslash \ell_\infty$ lies in $U_{\alpha\beta}$ if and only if $y - \alpha x^2 - \beta x^{q+1}$ is in GF$(q)$, which in turn is equivalent to

$$-a_1(x_0^2 + wx_1^2) - b_1(x_0^2 - wx_1^2) - 2a_0 x_0 x_1 + y_1 = 0. \quad (4.2)$$

It should be noted that we used $\epsilon^q = -\epsilon$ and $\epsilon^2 = w$ in the above computation. Thinking of $(x_0, x_1, y_0, y_1, 1)$ as the affine point of PG$(4, q) \backslash \Sigma_\infty$ corresponding to $(x, y, 1)$, we see that (4.2) is of the form $f(X_0, X_1) + X_3 X_4 = 0$, where $f$ is some binary quadratic form and $X_0 = x_0$, $X_1 = x_1$, $X_3 = y_1$, $X_4 = 1$. Moreover, as described in Section 1.2, $f$ is irreducible if and only if $(2a_0)^2 - 4(a_1 + b_1)(a_1 - b_1)w$ is a nonsquare in GF$(q)$. This expression can be rewritten as $(\beta^q - \beta)^2 + 4\alpha^{q+1}$. For the remainder of this section, we will let $d$ denote the expression $(\beta^q - \beta)^2 + 4\alpha^{q+1}$ in GF$(q)$, and think of it as being the *discriminant* of $U_{\alpha\beta}$.

The above argument shows that whenever $d$ is a nonsquare in GF$(q)$, the set of points $U_{\alpha\beta}$ defined in (4.1) constitutes an orthogonal-Buekenhout-Metz unital in $PG(2, q^2)$ which is tangent to $\ell_\infty$ at the point $P_\infty = (0, 1, 0)$. Conversely, using equation (4.2), we see that one can solve for $\alpha$ and $\beta$, given any irreducible binary quadratic form $f(X_0, X_1)$ (adding any element of $GF(q)$ to the second coordinate $\beta$ yields an equivalent unital, as we shall soon see, and thus $b_0$ may be arbitrarily specified). Thus we have proved the following theorem.

**Theorem 4.6.** *Every orthogonal-Buekenhout-Metz unital of $PG(2, q^2)$, for odd $q$, is equivalent to one expressed as $U_{\alpha\beta}$ for some $\alpha, \beta \in GF(q^2)$, where the discriminant $d = (\beta^q - \beta)^2 + 4\alpha^{q+1}$ is a nonsquare in $GF(q)$. Conversely, every such $U_{\alpha\beta}$ is an orthogonal-Buekenhout-Metz unital in $PG(2, q^2)$.*

Thus, for odd $q$, the unitals described in Theorem 4.6 are the only unitals embedded in $PG(2, q^2)$ which arise from one of Buekenhout's constructions. As mentioned before, it is presently unknown if $PG(2, q^2)$ contains any unitals which do not arise from one of Buekenhout's methods.

**Theorem 4.7.** *Let $U_{\alpha\beta}$ be an orthogonal-Buekenhout-Metz unital as defined in (4.1). Then $U_{\alpha\beta}$ is classical if $\alpha = 0$.*

*Proof.* Assume that $\alpha = 0$. Then from our discriminant condition we know that $(\beta^q - \beta)^2$ must be a nonsquare in $GF(q)$, and thus $\beta^q \neq \beta$. Using the coordinate approach discussed prior to Theorem 4.6, we see that the equation for $U_{0\beta}$ is $\beta x^{q+1} - y \in GF(q)$ and hence $\beta x^{q+1} - y = \beta^q x^{q+1} - y^q$. Using the fact that either $z = 1$ or $x = 0 = z$ for (normalized) points of $U_{0\beta}$, this equation can be rewritten as $(\beta - \beta^q)x^{q+1} + y^q z - yz^q = 0$. Multiplying by the nonzero field element $\epsilon$, we obtain the equivalent equation

$$\epsilon(\beta - \beta^q)x^{q+1} - \epsilon y z^q + \epsilon y^q z = 0.$$

Since $\epsilon^q = -\epsilon$ and $(\epsilon(\beta - \beta^q))^q = \epsilon(\beta - \beta^q) \neq 0$, this equation represents a nondegenerate Hermitian curve in $PG(2, q^2)$, and thus $U_{0\beta}$ is a classical unital. □

We shall soon see that $U_{\alpha\beta}$ is classical if and only if $\alpha = 0$. First we observe one can directly show that $U_{\alpha\beta}$ is a unital without resorting to Buekenhout's construction.

**Theorem 4.8.** *Let $q$ be an odd prime power, and let $\alpha, \beta \in GF(q^2)$ be such that $d = (\beta^q - \beta)^2 + 4\alpha^{q+1}$ is a nonsquare in $GF(q)$. Then*

$$U_{\alpha\beta} = \{(x, \alpha x^2 + \beta x^{q+1} + r) \mid x \in GF(q^2), r \in GF(q)\} \cup \{(0, 1, 0)\}$$

*is a unital in $PG(2, q^2)$.*

*Proof.* It suffices to show that every line of PG$(2, q^2)$ meets $U_{\alpha\beta}$ in 1 or $q + 1$ points. Using dual homogeneous coordinates, consider the line $(1, 0, \gamma)'$ for some $\gamma \in$ GF$(q^2)$. This line meets $U_{\alpha\beta}$ in the point set

$$\{(0, 1, 0)\} \cup \left\{ \left( -\gamma, \alpha\gamma^2 + \beta\gamma^{q+1} + r, 1 \right) \mid r \in \text{GF}(q) \right\}$$

and hence in $q + 1$ points. Also, the line $(0, 0, 1)'$ meets $U_{\alpha\beta}$ in the point set $\{(0, 1, 0)\}$ and hence in 1 point. All other lines look like $\ell = (\gamma, 1, \delta)'$ for some $\gamma, \delta \in$ GF$(q^2)$. Note that $(0, 1, 0)$ does not lie on $\ell$. Thus counting the number of points of $U_{\alpha\beta}$ lying on $\ell$ is the same as counting the number of elements $x$ in GF$(q^2)$ satisfying $\delta + \gamma x + \alpha x^2 + \beta x^{q+1} \in$ GF$(q)$. That is, we must count the number of solutions to the equation

$$\delta + \gamma x + \alpha x^2 + \beta x^{q+1} = \delta^q + \gamma^q x^q + \alpha^q x^{2q} + \beta^q x^{q+1}$$

in the variable $x$.

Choosing a basis $\{1, \epsilon\}$ for GF$(q^2)$ over GF$(q)$ as above and expanding with respect to this basis, we obtain the equation

$$(a_1 + b_1)x_0^2 + 2a_0 x_0 x_1 + (a_1 - b_1)wx_1^2 + g_1 x_0 + g_0 x_1 + d_1 = 0,$$

where $\gamma = g_0 + g_1 \epsilon$, and so on. This equation represents an affine conic in AG$(2, q)$. Completing AG$(2, q)$ to PG$(2, q)$ by adding the line at infinity with equation $x_2 = 0$, this affine conic gets completed to the projective conic $\mathcal{C}$ with equation

$$(a_1 + b_1)x_0^2 + 2a_0 x_0 x_1 + (a_1 - b_1)wx_1^2 + g_1 x_0 x_2 + g_0 x_1 x_2 + d_1 x_2^2 = 0.$$

Next note that $d = (\beta^q - \beta)^2 + 4\alpha^{q+1} = 4a_0^2 + 4(b_1^2 - a_1^2)w$, using the same basis expansion. By assumption, this expression must be a nonsquare in GF$(q)$. To determine how the conic $\mathcal{C}$ meets the line at infinity, we must solve the equation

$$(a_1 + b_1)x_0^2 + 2a_0 x_0 x_1 + (a_1 - b_1)wx_1^2 = 0.$$

Since $d$ is a nonsquare in GF$(q)$, the usual discriminant argument shows that this equation has only the trivial solution $x_0 = x_1 = 0$. Thus the conic $\mathcal{C}$ has no points on the line at infinity, and the number of points on the conic is the same as the number of points on $\ell \cap U_{\alpha\beta}$ as above. However, since the conic $\mathcal{C}$ (possibly degenerate) has no points on the line at infinity, $\mathcal{C}$ must either be an irreducible conic with all affine points (an "ellipse") or a single (affine) point. Hence $\ell$ meets $U_{\alpha\beta}$ in 1 or $q + 1$ points, proving the result. $\qquad\square$

We now compute the stabilizer of $U_{\alpha\beta}$, first in PGL$(3, q^2)$ and then in PΓL$(3, q^2)$. For each $\gamma \in$ GF$(q^2)$, let $\psi_\gamma$ denote the homography of PG$(2, q^2)$ given by

$$\psi_\gamma \colon (x,y,z) \mapsto (x + \gamma z, (2\alpha\gamma - (\beta^q - \beta)\gamma^q)x + y + (\alpha\gamma^2 + \beta\gamma^{q+1})z, z).$$

Similarly, for each $t \in \mathrm{GF}(q)$, let $\phi_t$ denote the homography of $\mathrm{PG}(2,q^2)$ given by

$$\phi_t \colon (x,y,z) \mapsto (x, y + tz, z).$$

Direct computations show that each $\phi_t$ and each $\psi_\gamma$ fixes the point $P_\infty = (0,1,0)$ and leaves invariant the unital $U_{\alpha\beta}$. More straightforward computations show that $\phi_t$ and $\psi_\gamma$ commute, as do $\phi_{t_1}$ and $\phi_{t_2}$. Moreover,

$$\phi_{t_1}\phi_{t_2} = \phi_{t_1+t_2},$$
$$\psi_{\gamma_1}\psi_{\gamma_2} = \psi_{\gamma_1+\gamma_2}\phi_{-(\beta\gamma_1^q\gamma_2 + \beta^q\gamma_1\gamma_2^q)},$$
$$(\psi_{\gamma_1}\phi_{t_1})(\psi_{\gamma_2}\phi_{t_2}) = \psi_{\gamma_1+\gamma_2}\phi_{t_1+t_2-(\beta\gamma_1^q\gamma_2 + \beta^q\gamma_1\gamma_2^q)}.$$

Thus $S = \{\psi_\gamma\phi_t \mid \gamma \in \mathrm{GF}(q^2), t \in \mathrm{GF}(q)\}$ is a homography subgroup of order $q^3$ that fixes $P_\infty$ and stabilizes $U_{\alpha\beta}$; it is abelian if and only if $\beta^q - \beta = 0$ (equivalently, $\beta \in \mathrm{GF}(q)$). Also, $K = \{\phi_t \mid t \in \mathrm{GF}(q)\}$ is an elementary abelian $p$-group (that is, every nonidentity element has order $p$) of order $q$, where $q$ is a power of the prime $p$, that fixes $P_\infty$ and stabilizes $U_{\alpha\beta}$. In fact, $K$ fixes each point on $\ell_\infty$ and each line through $P_\infty$. That is, $K$ is an elation subgroup of order $q$ with center $P_\infty$ and axis $\ell_\infty$. In particular, $K$ acts transitively on the points, other than $P_\infty$, of any chord of the unital through $P_\infty$. Finally, since the stabilizer of any point of $U_{\alpha\beta} \setminus P_\infty$ in the subgroup $S$ consists only of the identity collineation, the Orbit-Stabilizer Theorem implies that $S$ acts transitively on $U_{\alpha\beta} \setminus P_\infty$.

**Lemma 4.9.** *Let $\delta$ be a nonzero element of $\mathrm{GF}(q^2)$, and let $\mu_\delta$ denote the homography of $\mathrm{PG}(2,q^2)$ given by*

$$\mu_\delta \colon (x,y,z) \mapsto (\delta x, \delta^2 y, z).$$

*If $\beta \notin \mathrm{GF}(q)$, then $\mu_\delta$ leaves $U_{\alpha\beta}$ invariant if and only if $\delta \in \mathrm{GF}(q)$. If $\beta \in \mathrm{GF}(q)$, then $\mu_\delta$ leaves $U_{\alpha\beta}$ invariant if and only if $\delta$ is a power of $\epsilon$.*

*Proof.* First note that $\mu_\delta$ fixes the point $P_\infty$. Let $(x, \alpha x^2 + \beta x^{q+1} + r, 1)$ be a generic point of $U_{\alpha\beta}$. Its image under $\mu_\delta$ is $(\delta x, \delta^2\alpha x^2 + \delta^2\beta x^{q+1} + \delta^2 r, 1)$. Hence $\mu_\delta$ leaves $U_{\alpha\beta}$ invariant if and only if $(\delta^{q+1} - \delta^2)\beta x^{q+1} - \delta^2 r \in \mathrm{GF}(q)$ for all $x \in \mathrm{GF}(q^2)$ and for all $r \in \mathrm{GF}(q)$. Taking $x = 0$ and $r = 1$, we see that necessarily $\delta^2 \in \mathrm{GF}(q)$. Thus, since $x^{q+1}$ is an element of $\mathrm{GF}(q)$ for all $x \in \mathrm{GF}(q^2)$, a necessary and sufficient condition for $\mu_\delta$ to leave $U_{\alpha\beta}$ invariant is that both $\beta(\delta^{q+1} - \delta^2)$ and $\delta^2$ belong to $\mathrm{GF}(q)$. One should also note that if $\delta^2$ is in $\mathrm{GF}(q)$, then so is $\delta^{q+1}$ since $q$ is odd and hence $q + 1$ is even.

If $\beta \notin \mathrm{GF}(q)$, then the necessary and sufficient condition becomes $\delta^2 = \delta^{q+1}$; that is, $\delta^{q-1} = 1$ and hence $\delta \in \mathrm{GF}(q)$. If $\beta \in \mathrm{GF}(q)$, then the necessary and sufficient condition reduces to $\delta^2 \in \mathrm{GF}(q)$ as above, or equivalently $\delta^{2(q-1)} = 1$. Since the multiplicative order of $\epsilon$ is $2(q - 1)$, this latter condition simply says that $\delta$ must be a power of $\epsilon$. $\square$

**Theorem 4.10.** *Let $J$ be the cyclic subgroup generated by $\mu_\delta$, where $\delta = \epsilon$ if $\beta \in GF(q)$ and $\delta = \epsilon^2 = w$ if $\beta \notin GF(q)$. Then $J$ is a homography subgroup of order $2(q - 1)$ or $q - 1$, respectively, that fixes $P_\infty$ and leaves $U_{\alpha\beta}$ invariant. Moreover, if $S$ is the subgroup of order $q^3$ discussed above, then $J$ normalizes $S$ and $J$ meets $S$ only in the identity collineation.*

*Proof.* This follows immediately from the above lemma and the direct computation showing that $\mu_\delta^{-1} \psi_\gamma \phi_t \mu_\delta = \psi_{\gamma\delta} \phi_{t\delta^2}$.     □

**Corollary 4.11.** *The semidirect product $H$ of $S$ by the cyclic group $J$ is a homography subgroup of order $q^3(q - 1)$ or $2q^3(q - 1)$, according to $\beta \notin GF(q)$ or $\beta \in GF(q)$, that fixes $P_\infty$ and acts transitively on $U_{\alpha\beta} \setminus P_\infty$. The subgroup $S$ is a Sylow p-subgroup of $H$, where $q$ is a power of the odd prime $p$, which is abelian if and only if $\beta \in GF(q)$. Moreover, the subgroup $K$ of $S$ is an elation subgroup of order $q$ with axis $\ell_\infty$ and center $P_\infty$ that acts transitively on the points, other than $P_\infty$, of each chord through $P_\infty$.*

We now let $G$ be the subgroup of $P\Gamma L(3, q^2)$ that fixes $P_\infty = (0, 1, 0)$ and leaves the orthogonal-Buekenhout-Metz unital $U_{\alpha\beta}$ invariant. Since the unique tangent line to $U_{\alpha\beta}$ at $P_\infty$ is $\ell_\infty = (0, 0, 1)'$, $G$ must also leave this line invariant. Thus each collineation in $G$ is of the form

$$(x, y, z) \mapsto (\theta_1 x^\tau + \theta_2 z^\tau, \ \theta_3 x^\tau + \theta_4 y^\tau + \theta_5 z^\tau, \ z^\tau),$$

where $\theta_1, \theta_2, \theta_3, \theta_4, \theta_5 \in GF(q^2)$ with $\theta_1\theta_4 \neq 0$, and $\tau$ is some field automorphism of $GF(q^2)$. Careful consideration of the conditions imposed on the parameters of this mapping by the stabilization of $U_{\alpha\beta}$ enables one to explicitly compute the group $G$. The following theorem, whose proof can be found in [22], summarizes the situation.

**Theorem 4.12.** *Let $U_{\alpha\beta}$ be an orthogonal-Buekenhout-Metz unital in PG$(2, q^2)$, where $q = p^e$ is an odd prime power. Let $F$ denote $GF(p)$ if $\alpha = 0$ or the smallest subfield of $GF(q)$ containing the element $d' = (\beta^q - \beta)^2/4\alpha^{q+1}$ if $\alpha \neq 0$. Then*

$$o(G) = \begin{cases} mq^3(q^2 - 1) & \text{if } \alpha = 0 \\ 2mq^3(q - 1) & \text{if } \beta \in GF(q) \\ mq^3(q - 1) & \text{if } \alpha \neq 0 \text{ and } \beta \notin GF(q), \end{cases}$$

*where $m$ is the dimension of $GF(q^2)$ over the subfield $F$. Moreover, the homography subgroup of $G$, namely, $G_0 = G \cap PGL(3, q^2)$, has index $m$ in $G$. The group $G$ acts transitively on the points of $U_{\alpha\beta} \setminus P_\infty$, acts transitively on the points of $\ell_\infty \setminus P_\infty$, and has one or two orbits on the points of PG$(2, q^2) \setminus (U_{\alpha\beta} \cup \ell_\infty)$.*

**Corollary 4.13.** *The orthogonal-Buekenhout-Metz unital $U_{\alpha\beta}$ is classical if and only if $\alpha = 0$.*

*Proof.* We have already shown than $U_{\alpha\beta}$ is classical if $\alpha = 0$. The converse follows from the above theorem and the known order for a point stabilizer in $\mathrm{PGU}(3, q^2)$.                                                                  □

It turns out that using some nontrivial results from group theory (see [22] for the details) one can show that if $U_{\alpha\beta}$ is nonclassical, then any collineation leaving $U_{\alpha\beta}$ invariant must fix the point $P_\infty$. Hence, if $\alpha \neq 0$, the groups $G$ and $G_0$ of Theorem 4.12 are the full stabilizers of $U_{\alpha\beta}$ in $\mathrm{P\Gamma L}(3, q^2)$ and $\mathrm{PGL}(3, q^2)$, respectively. Note also that $G_0$ is precisely the group $H$ described in Corollary 4.11 when $\alpha \neq 0$.

We now discuss equivalence among the orthogonal-Buekenhout-Metz unitals $U_{\alpha\beta}$, as previously promised. Since $U_{\alpha\beta}$ is classical if and only if $\alpha = 0$ (by Corollary 4.13) and since any two classical unitals are equivalent, we may assume that $\alpha \neq 0$, and thus the point $P_\infty$ is fixed by any collineation stabilizing $U_{\alpha\beta}$. Hence, if $\zeta : U_{\alpha\beta} \mapsto U_{\alpha'\beta'}$, we can assume that $\zeta$ fixes $P_\infty$ and thus $\zeta$ leaves invariant the line $\ell_\infty$. Since the stabilizer of $U_{\alpha'\beta'}$ fixes $P_\infty$ and $\ell_\infty$ while acting transitively on $U_{\alpha'\beta'} \setminus P_\infty$, we can further assume without loss of generality that $\zeta$ maps the point $Q = (0, 0, 1)$ to itself. Hence the collineation $\zeta$ is of the form

$$\zeta : (x, y, z) \mapsto (\kappa x^\tau,\ \lambda x^\tau + \nu y^\tau,\ z^\tau),$$

where $\kappa, \lambda, \nu \in \mathrm{GF}(q^2)$ with $\kappa\nu \neq 0$ and $\tau$ is some field automorphism of $\mathrm{GF}(q^2)$.

Since $(0, 1, 1) \in U_{\alpha\beta}$, we have $(0, 1, 1)\zeta = (0, \nu, 1) \in U_{\alpha'\beta'}$ and thus $\nu \in \mathrm{GF}(q)$. Similarly, if $(x, y, 1) \in U_{\alpha\beta}$, we have

$$(x, y, 1)\zeta = (\kappa x^\tau, \lambda x^\tau + \nu y^\tau, 1) \in U_{\alpha'\beta'}$$

and hence $\alpha'(\kappa x^\tau)^2 + \beta'(\kappa x^\tau)^{q+1} - \lambda x^\tau - \nu y^\tau \in \mathrm{GF}(q)$. Since $\nu \in \mathrm{GF}(q)$ and thus $\nu(\alpha x^2 + \beta x^{q+1} - y)^\tau \in \mathrm{GF}(q)$, we have

$$(\alpha'\kappa^2 - \alpha^\tau\nu)(x^\tau)^2 + (\beta'\kappa^{q+1} - \beta^\tau\nu)(x^\tau)^{q+1} - \lambda x^\tau \in \mathrm{GF}(q).$$

Successively setting $x^\tau$ equal to $1, -1, \epsilon, -\epsilon$, and $1 + \epsilon$ yields a system of equations which implies that

$$\alpha'\kappa^2 - \alpha^\tau\nu = 0,$$
$$\beta'\kappa^{q+1} - \beta^\tau\nu \in \mathrm{GF}(q),$$
$$\lambda = 0.$$

Therefore, if $U_{\alpha'\beta'}$ is to be projectively equivalent to $U_{\alpha\beta}$, we must have

$$(\alpha', \beta') = (\alpha^\tau\gamma^2\nu, \beta^\tau\gamma^{q+1}\nu + u), \tag{4.3}$$

where $\nu$ is a nonzero element of $\mathrm{GF}(q)$, $\gamma$ is a nonzero element of $\mathrm{GF}(q^2)$, $u$ is some element of $\mathrm{GF}(q)$, and $\tau$ is some field automorphism of $\mathrm{GF}(q^2)$.

Conversely, straightforward computations show that these conditions are sufficient. We will write $(\alpha, \beta) \sim (\alpha', \beta')$ in this case, and note that $\sim$ is an equivalence relation on the ordered pairs $(\alpha, \beta)$ with discriminant $d = (\beta^q - \beta)^2 + 4\alpha^{q+1}$ a nonsquare in GF$(q)$.

**Theorem 4.14.** *Let $q = p^e$ be an odd prime power, and write $e = 2^t e_0$, where $e_0$ is odd and $t$ is a nonnegative integer. Then the number of mutually inequivalent orthogonal-Buekenhout-Metz unitals of order $q$ in* PG$(2, q^2)$ *is*

$$\frac{1}{2e} \left[ e_0 + \sum_{m|e} \varphi\left(\frac{2e}{m}\right) p^m \right],$$

*where $\varphi$ is the Euler totient function.*

*Proof.* Using the above notation, we write $(\alpha, \beta) \sim (\alpha', \beta')$ to denote that $U_{\alpha\beta}$ is equivalent to $U_{\alpha'\beta'}$. As discussed previously, there is one equivalence class when $\alpha = 0$, namely, the class of nondegenerate Hermitian curves. Similarly, we can show there is one equivalence class when $\beta \in$ GF$(q)$. Namely, consider $(\alpha, \beta)$ and $(\alpha', \beta')$ with $\beta, \beta' \in$ GF$(q)$. Then $\alpha$ and $\alpha'$ are nonsquares in GF$(q^2)$ from the discriminant condition, and thus $\alpha'/\alpha = \gamma^2$ for some nonzero element $\gamma \in$ GF$(q^2)$. Since $\gamma^{q+1} \in$ GF$(q)$, we may further take $\tau = id, v = 1$, and $u = \beta' - \beta\gamma^{q+1}$ to see that $(\alpha, \beta) \sim (\alpha', \beta')$ according to condition (4.3).

We now assume that $\alpha \neq 0$ and $\beta \notin$ GF$(q)$. Thus we may write $\beta = r + s\zeta$, where $r, s \in$ GF$(q)$ and $s \neq 0$. Using the norm from GF$(q^2)$ to GF$(q)$, as discussed in Section 1.2, there is some $\gamma \in$ GF$(q^2)$ such that $\gamma^{q+1} = s^{-1}$ and hence $(\alpha, \beta) \sim (\alpha\gamma^2, \beta\gamma^{q+1} - r\gamma^{q+1}) = (\alpha\gamma^2, \zeta)$ from condition (4.3). That is, without loss of generality, we may assume that $\beta = \zeta$, where $\zeta$ is a primitive element of GF$(q^2)$ as previously defined.

Let $d' = (\beta^q - \beta)^2/4\alpha^{q+1}$ as defined in Theorem 4.12. Since $(\beta^q - \beta)^2$ is a nonsquare in GF$(q)$, as is the discriminant $d = (\beta^q - \beta)^2 + 4\alpha^{q+1}$, it follows that $d'(d' + 1) = (\beta^q - \beta)^2 d/16\alpha^{2(q+1)}$ is a nonzero square in GF$(q)$. Moreover, given any $\tilde{d} \in$ GF$(q)$ with $\tilde{d}(\tilde{d} + 1)$ a nonzero square in GF$(q)$, by thinking of norms once again, one can choose (in $q + 1$ ways) an element $\tilde{\alpha} \in$ GF$(q^2)$ such that $\tilde{\alpha}^{q+1} = (\beta^q - \beta)^2/4\tilde{d}$ and then check that $(\beta^q - \beta)^2 + 4\tilde{\alpha}^{q+1}$ is a nonsquare in GF$(q)$. Thus the resulting $U_{\tilde{\alpha}\beta}$ is an orthogonal-Buekenhout-Metz unital.

Now let $d', \tilde{d} \in$ GF$(q)$ with $d'(d' + 1)$ and $\tilde{d}(\tilde{d} + 1)$ nonzero squares in GF$(q)$. Choose (in any way) $\alpha, \tilde{\alpha} \in$ GF$(q^2)$ such that $\alpha^{q+1} = (\beta^q - \beta)^2/4d'$ and $\tilde{\alpha}^{q+1} = (\beta^q - \beta)^2/4\tilde{d}$. Then routine computations (see [22] if needed) show that $(\alpha, \beta) \sim (\tilde{\alpha}, \beta)$ if and only if $\tilde{d} = (d')^\tau$ for some field automorphism $\tau$ of GF$(q^2)$. In this case we write $d' \approx \tilde{d}$ and observe that $\approx$ is another equivalence relation. Hence the number of mutually inequivalent orthogonal-Buekenhout-Metz unitals in PG$(2, q^2)$ is $N + 2$, where $N$ is the number of equivalence classes under $\approx$ on the elements $d' \in GF(q)$ such that $d'(d'+1)$ is a nonzero square in $GF(q)$.

Recalling that $q = p^e$, for each positive divisor $m$ of $e$ we define $N_m$ to be the number of elements $d' \in GF(p^m)$ such that $d'(d'+1)$ is a nonzero square in $GF(q)$ and $d'$ is contained in no smaller subfield of $GF(q)$. Since the automorphism group of $GF(p^m)$ has order $m$ and field automorphisms preserve subfields as well as quadratic character (squares and nonsquares), we see that $N = \sum_{m|e} N_m/m$. Now, if $e/m$ is even, then every element of $GF(p^m)$ is a (possibly zero) square in $GF(q)$ and thus, excluding the elements $0$ and $-1$, we have $\sum_{m'|m} N_{m'} = p^m - 2$. On the other hand, if $e/m$ is odd, $d'(d'+1)$ is a nonzero square in $GF(q)$ precisely when $d'(d'+1)$ is a nonzero square in $GF(p^m)$. That is, we need to count the number of elements $d' \in GF(p^m)$ such that $d'$ and $d'+1$ are both nonsquares or both nonzero squares in $GF(p^m)$. By an elementary cyclotomy result (see [98], for instance), this number is always $(p^m - 3)/2$, and so $\sum_{m'|m} N_{m'} = (p^m - 3)/2$ in this case. Hence, both cases considered, we have

$$\sum_{m'|m} N_{m'} = \frac{1}{2}\left[(p^m - 1)\left(\frac{3}{2} + \frac{1}{2}(-1)^{\frac{e}{m}}\right)\right] - 1.$$

The result now follows from a tedious application of Mobiüs inversion.    □

**Corollary 4.15.** *If $q$ is an odd prime, then the number of mutually inequivalent orthogonal-Buekenhout-Metz unitals in $PG(2, q^2)$ is $(q+1)/2$, one of which is the Hermitian curve.*

Next we discuss the lines in $PG(2, q^2)$ which are tangent to the unital $U_{\alpha\beta}$. Our goal is to show that $U_{\alpha\beta}$ is a self-dual unital for all choices of $\alpha$ and $\beta$. From definition (4.1) one sees that the tangent line to $U_{\alpha\beta}$ at the point $P_\infty$ is $\ell_\infty = (0, 0, 1)$. To compute the other tangent lines it is best to use our transitivity results.

**Lemma 4.16.** *Let $q$ be an odd prime power. Then the tangent line to $U_{\alpha\beta}$ at the point $(x, \alpha x^2 + \beta x^{q+1} + r, 1)$ is $(-2\alpha x + (\beta^q - \beta)x^q, 1, \alpha x^2 - \beta^q x^{q+1} - r)'$.*

*Proof.* We first claim that $\ell = (0, 1, 0)'$ is the tangent line to $U_{\alpha\beta}$ at the point $P = (0, 0, 1)$. Clearly $P \in \ell$, and any other point of $\ell \cap U_{\alpha\beta}$ looks like $(x, 0, 1)$ for some $x \in GF(q^2)$. But then (4.1) implies that $\alpha x^2 + \beta x^{q+1} \in GF(q)$, and therefore $\alpha x^2 + \beta x^{q+1} = \alpha^q x^{2q} + \beta^q x^{q+1}$. This implies that

$$(\beta^q - \beta)^2 x^{2(q+1)} = (\alpha x^2 - \alpha^q x^{2q})^2$$

and thus $[(\beta^q - \beta)^2 + 4\alpha^{q+1}]x^{2(q+1)} = (\alpha x^2 + \alpha^q x^{2q})^2$. Since $x^{2(q+1)}$ and $(\alpha x^2 + \alpha^q x^{2q})^2$ are (possibly zero) squares in $GF(q)$, while the discriminant $d = (\beta^q - \beta)^2 + 4\alpha^{q+1}$ is a nonsquare in $GF(q)$, necessarily $x = 0$ and thus $\ell = (0, 1, 0)'$ is the unique tangent line to $U_{\alpha\beta}$ at $P = (0, 0, 1)$.

Using the collineations defined prior to Lemma 4.9, we see that $\psi_x \phi_r$ maps the point $(0, 0, 1)$ to the point $(x, \alpha x^2 + \beta x^{q+1} + r, 1)$. Thus, using dual homogeneous coordinates for lines,

$$(\psi_x \phi_r)^{-1} = \psi_{-x} \phi_{-(\beta^q + \beta) x^{q+1} - r}$$

maps $(0, 1, 0)'$ to the line coordinates for the (unique) tangent line to $U_{\alpha\beta}$ at $(x, \alpha x^2 + \beta x^{q+1} + r, 1)$. The result now follows by a direct computation. □

**Theorem 4.17.** *Let $q$ be an odd prime power. Then the orthogonal-Buekenhout-Metz unital $U_{\alpha\beta}$ in PG$(2, q^2)$ defined by (4.1) is self-dual.*

*Proof.* As shown in Lemma 4.16, the tangent line at $(x, \alpha x^2 + \beta x^{q+1} + r, 1)$ is $(-2\alpha x + (\beta^q - \beta) x^q, 1, \alpha x^2 - \beta^q x^{q+1} - r)'$, while the tangent line at $(0, 1, 0)$ is $(0, 0, 1)'$. Letting $\tilde{x} = -2\alpha x + (\beta - \beta) x^q$ and $\tilde{y} = \alpha x^2 - \beta^q x^{q+1} - r$, straightforward computations show that $d\tilde{y} - \alpha^q \tilde{x}^2 - \beta^q \tilde{x}^{q+1} \in GF(q)$, where $d = (\beta^q - \beta)^2 + 4\alpha^{q+1}$ is the discriminant of $U_{\alpha\beta}$. Thus the dual coordinates for the tangent lines at the affine points of $U_{\alpha\beta}$ satisfy the "equation"

$$\tilde{y} - \frac{\alpha^q}{d} \tilde{x}^2 - \frac{\beta^q}{d} \tilde{x}^{q+1} \in GF(q).$$

Also, the "discriminant" of $U_{\frac{\alpha^q}{d} \frac{\beta^q}{d}}$ is $\frac{1}{d}$, which is again a nonsquare in GF$(q)$. Hence, interchanging second and third coordinates, one sees that the dual unital to $U_{\alpha\beta}$ is isomorphic to the orthogonal-Buekenhout-Metz unital $U_{\frac{\alpha^q}{d} \frac{\beta^q}{d}}$.

Moreover, using condition (4.3), we have

$$\left( \frac{\alpha^q}{d}, \frac{\beta^q}{d} \right) \sim (\alpha^q, \beta^q) \sim \left( \alpha^q \alpha^2 \alpha^{-(q+1)}, \beta^q \alpha^{q+1} \alpha^{-(q+1)} \right)$$
$$= (\alpha, \beta^q) \sim (\alpha, \beta^q - (\beta + \beta^q))$$
$$= (\alpha, -\beta) \sim \left( \alpha \epsilon^2 w^{-1}, \beta \epsilon^{q+1} w^{-1} \right) = (\alpha, \beta).$$

Therefore $U_{\alpha\beta}$ is self-dual. □

We now discuss the feet, as previously defined, of some point $Q$ in PG$(2, q^2) \setminus U_{\alpha\beta}$. That is, we look at the $q + 1$ points of $U_{\alpha\beta}$ that lie on the $q + 1$ tangent lines through $Q$. In the classical case, these feet are always collinear, lying on $Q^\perp$. In the nonclassical case, $U_{\alpha\beta}$ does not consist of the absolute points of some polarity of PG$(2, q^2)$ and hence the feet may or may not be collinear.

**Theorem 4.18.** *Let $U_{\alpha\beta}$ be an orthogonal-Buekenhout-Metz unital of PG$(2, q^2)$, for some odd prime power $q$, which is not classical. Let $Q \in$ PG$(2, q^2) \setminus U_{\alpha\beta}$. Then the feet of $Q$ are collinear if and only if $Q \in \ell_\infty$.*

*Proof.* Since $U_{\alpha\beta}$ is nonclassical, $\alpha \neq 0$ by Corollary 4.13. Suppose first that $Q \in \ell_\infty$. Then $Q = (1, \gamma, 0)$ for some $\gamma \in \mathrm{GF}(q^2)$. According to Lemma 4.16, every tangent line to $U_{\alpha\beta}$, other than $\ell_\infty$, is of the form

$$(-2\alpha x + (\beta^q - \beta)x^q, 1, \alpha x^2 - \beta^q x^{q+1} - r)',$$

for some $x \in \mathrm{GF}(q^2)$ and some $r \in \mathrm{GF}(q)$. Thus the tangent lines through $Q$, other than $\ell_\infty$, have dual coordinates that satisfy the equation

$$(\beta^q - \beta)x^q - 2\alpha x + \gamma = 0.$$

Expanding with respect to the basis $\{1, \epsilon\}$ for $\mathrm{GF}(q^2)$ over $\mathrm{GF}(q)$, as before, we obtain the two equations

$$2a_0 x_0 + 2(a_1 - b_1)w x_1 = g_0$$

and

$$2(a_1 + b_1)x_0 + 2a_0 x_1 = g_1,$$

where $\gamma = g_0 + g_1 \epsilon$, and so on. The determinant of this linear system is the discriminant $d$ of $U_{\alpha\beta}$, which is nonzero. Hence there is a unique solution $(x_0, x_1)$ and thus a unique solution for $x$, say $x = \theta$. Hence the feet of $Q$ are

$$\{(\theta, \alpha\theta^2 + \beta\theta^{q+1} + r, 1) \mid r \in \mathrm{GF}(q)\} \cup \{(0, 1, 0)\},$$

all of which lie on the line $(1, 0, -\theta)'$. Note that this holds true independent of the value of $\alpha$.

Now suppose that $Q \notin \ell_\infty$. According to Theorem 4.12, the stabilizer of $U_{\alpha\beta}$ has at most two orbits on $\mathrm{PG}(2, q^2)\setminus(\ell_\infty \cup U_{\alpha\beta})$. In fact, representatives for these two orbits, which may or may not be combined depending upon $\alpha$ and $\beta$, are $(0, \epsilon, 1)$ and $(0, w\epsilon, 1)$, where $w = \epsilon^2$ is a primitive element of $\mathrm{GF}(q)$ as previously defined. We only consider $Q = (0, \epsilon, 1)$, the other case being completely analogous.

By Lemma 4.16, the $q + 1$ tangent lines through $Q$ have dual homogeneous coordinates of the form $(-2\alpha x + (\beta^q - \beta)x^q, 1, \alpha x^2 - \beta^q x^{q+1} - r)'$, where $x \in \mathrm{GF}(q^2)$ and $r = \epsilon + \alpha x^2 - \beta^q x^{q+1} \in \mathrm{GF}(q)$. The associated point of $U_{\alpha\beta}$ on such a tangent line is

$$(x, \alpha x^2 + \beta x^{q+1} + r, 1) = (x, 2\alpha x^2 + (\beta - \beta^q)x^{q+1} + \epsilon, 1).$$

Expanding with respect to the basis $\{1, \epsilon\}$ for $\mathrm{GF}(q^2)$ over $\mathrm{GF}(q)$ as in the first part of the argument, we see that the feet of $Q$ are the sets of points

$$(x_0 + x_1 \epsilon, 2a_0 x_0^2 + 2a_0 w x_1^2 + 4a_1 w x_0 x_1 - \epsilon, 1)$$

satisfying the condition

$$(a_1 + b_1)x_0^2 + (a_1 - b_1)w x_1^2 + 2a_0 x_0 x_1 + 1 = 0.$$

Note that the argument given in the proof of Theorem 4.8 shows that there are indeed $q + 1$ feet, as the above condition on the feet is the equation of an ellipse in AG$(2, q)$.

Now suppose that these feet are collinear, say all lying on a line with dual homogeneous coordinates $(s_0 + s_1\epsilon, 1, t_0 + t_1\epsilon)'$. Then we must have

$$2a_0 x_0^2 + 2a_0 w x_1^2 + 4a_1 w x_0 x_1 + s_0 x_0 + s_1 w x_1 + t_0 = 0,$$

$$s_1 x_0 + s_0 x_1 + t_1 - 1 = 0$$

satisfied for all the feet. The first equation above again represents some affine conic in AG$(2, q)$. Hence this conic must be identical to the ellipse whose equation defines the $X$-coordinates for the feet. This forces $s_0 = 0 = s_1$ and thus $t_1 = 1$ from the second equation. Comparing the resulting two equations for the defining ellipse, we see that necessarily $a_0 t_0 = 2w a_1$ and $a_1 t_0 = 2a_0$. Since $\alpha \neq 0$, the latter equations imply that $a_0 \neq 0$ and $a_1 \neq 0$, and therefore $2a_0/a_1 = t_0 = 2a_1 w/a_0$. Hence $a_0^2 - w a_1^2 = 0$ and thus $\alpha^{q+1} = 0$, contradicting the fact that $\alpha \neq 0$. Thus no such line contains the feet of $Q$.

The only other possibility is $(1, 0, t_0 + t_1\epsilon)'$ for a line containing the feet, in which case one must have $x_0 = -t_0$ and $x_1 = -t_1$. This contradicts the fact that there are $q + 1$ feet, not 1. So, for nonclassical unitals $U_{\alpha\beta}$ the feet of $Q$ are never collinear. The result now follows by a transitivity argument as indicated above.                                                                     □

We have observed that when $\beta \in$ GF$(q)$, and hence $\alpha \neq 0$ by the discriminant condition, the nonclassical unital $U_{\alpha\beta}$ has a slightly larger stabilizer than in the other nonclassical cases. Moreover, the Sylow $p$-subgroup $S$ of $G_0$ is abelian in this case. This leads one to believe that perhaps $U_{\alpha\beta}$ has some extra geometry associated with it when $\beta \in$ GF$(q)$. Indeed, this is the case.

**Theorem 4.19.** *Let $U_{\alpha\beta}$ be an orthogonal-Buekenhout-Metz unital in PG$(2, q^2)$, for some odd prime power $q$. Then $U_{\alpha\beta}$ contains no irreducible conic disjoint from the point $P_\infty = (0, 1, 0)$. Moreover, $U_{\alpha\beta}$ contains an irreducible conic passing through $P_\infty$ if and only if $\beta \in$ GF$(q)$, in which case $U_{\alpha\beta}$ is the union of $q$ irreducible conics pairwise meeting in the point $P_\infty$.*

*Proof.* Recall that irreducible conics of PG$(2, q^2)$ consist of $q^2 + 1$ points, no three of which are collinear. Assume that $\mathcal{C}$ is some irreducible conic of PG$(2, q^2)$ which is contained in $U_{\alpha\beta}$. Then the unique tangent line to $U_{\alpha\beta}$ at any point of $\mathcal{C}$ must also be the unique tangent line to $\mathcal{C}$ at that point.

Suppose first that $P_\infty = (0, 1, 0) \notin \mathcal{C}$. Since $q$ is odd, $P_\infty$ is either interior or exterior to $\mathcal{C}$, and hence $P_\infty$ lies on at least $(q^2 - 1)/2$ secants to $\mathcal{C}$. Using the transitivity described in Theorem 4.12, without loss of generality we may assume that $Q = (0, 0, 1)$ is a point of $\mathcal{C}$ on some secant line through $P_\infty$. The other point of $\mathcal{C} \subset U_{\alpha\beta}$ on this secant line $(1, 0, 0)'$ looks like $R = (0, r, 1)$, for some nonzero element $r$ in GF$(q)$. The tangent lines to $\mathcal{C}$ at $Q$ and $R$ are $(0, 1, 0)'$ and $(0, 1, -r)'$, respectively, by Lemma 4.16. This is enough information to force the equation of $\mathcal{C}$ to be of the form

$$\gamma X^2 - Y^2 + rYZ = 0,$$

for some nonzero $\gamma$ in $\mathrm{GF}(q^2)$.

Since $\ell_\infty = (0,0,1)'$ is a tangent line to $U_{\alpha\beta}$, it cannot be secant to $\mathcal{C}$. Thus the secant lines to $\mathcal{C}$ through $P_\infty = (0,1,0)$, other than $QR$, are all of the form $(1,0,-\delta)$, for some nonzero $\delta$ of $\mathrm{GF}(q^2)$. The $Y$-coordinates of the two points of $\mathcal{C} \subset U_{\alpha\beta}$ on any such secant must satisfy the equation

$$Y^2 - rY - \gamma\delta^2 = 0,$$

and hence the discriminant $r^2 + 4\gamma\delta^2$ of this quadratic equation must be a nonzero square in $\mathrm{GF}(q^2)$. Moreover, since the $X$-coordinates of these two points are the same, namely, $\delta$, the difference in the $Y$-coordinates of these two points of $U_{\alpha\beta}$ must be an element of $\mathrm{GF}(q)$. This difference is $\sqrt{r^2 + 4\gamma\delta^2}$ by the quadratic formula. Since $\gamma \neq 0$, there cannot be $(q^2 - 3)/2$ values of $\delta$ for which this holds. Hence $U_{\alpha\beta} \setminus P_\infty$ contains no irreducible conic.

Now suppose that an irreducible conic $\mathcal{C} \subset U_{\alpha\beta}$ contains the point $P_\infty = (0,1,0)$. Again using the transitivity described in Theorem 4.12, we may assume that the conic $\mathcal{C}$ contains the point $Q = (0,0,1)$. Then the tangent lines to $U_{\alpha\beta}$ at $P_\infty$ and $Q$, namely, $\ell_\infty = (0,0,1)'$ and $(0,1,0)'$, respectively, must be tangent to $\mathcal{C}$, and hence the equation of $\mathcal{C}$ must be of the form

$$\gamma X^2 - YZ = 0,$$

for some nonzero $\gamma$ in $\mathrm{GF}(q^2)$. Consider the $q^2$ points of $\mathcal{C} \subset U_{\alpha\beta}$, other than $P_\infty$. These points all have distinct $X$-coordinates, since otherwise $\mathcal{C}$ would contain three collinear points, one being $P_\infty$. In particular, this means every element of $\mathrm{GF}(q^2)$ appears as an $X$-coordinate for one of the points of $\mathcal{C} \setminus P_\infty$. Taking $X = 1$ and $X = \epsilon$ in turn, and using the above equation for $\mathcal{C}$ as well as definition (4.1) for $U_{\alpha\beta}$, we see that both $\gamma - \alpha - \beta$ and $(\gamma - \alpha)w + \beta w$ must be elements of $\mathrm{GF}(q)$. As $w$ is a nonzero element of $\mathrm{GF}(q)$, these conditions imply that $\beta$ must be an element of $\mathrm{GF}(q)$ whenever $U_{\alpha\beta}$ contains an irreducible conic passing through the point $P_\infty$.

Conversely, consider an orthogonal-Buekenhout-Metz unital $U_{\alpha\beta}$ with $\beta \in \mathrm{GF}(q)$. The discriminant condition for $U_{\alpha\beta}$ then implies that $4\alpha^{q+1}$ is a nonsquare in $\mathrm{GF}(q)$, and in particular, $\alpha \neq 0$. Since all such unitals are equivalent, as shown in the proof of Theorem 4.14, we may as well assume that $\beta = 0$, and hence

$$U_{\alpha 0} = \{(x, \alpha x^2 + r, 1) \mid x \in \mathrm{GF}(q^2), r \in \mathrm{GF}(q)\} \cup \{(0,1,0)\}.$$

Then $U_{\alpha 0} = \bigcup_{r \in \mathrm{GF}(q)} \mathcal{C}_r$, where

$$\mathcal{C}_r = \{(x, \alpha x^2 + r, 1) \mid x \in \mathrm{GF}(q^2)\} \cup \{(0,1,0)\}$$

is an irreducible conic in $\mathrm{PG}(2, q^2)$ with equation $\alpha X^2 + rZ^2 - YZ = 0$. Since any two of these conics meet in the point $P_\infty$, the proof is complete.    $\square$

**Corollary 4.20.** *The Hermitian curve in* PG$(2, q^2)$, *where $q$ is an odd prime power, contains no irreducible conic.*

*Proof.* By Corollary 4.13, any Hermitian curve in PG$(2, q^2)$ is equivalent to $U_{0\beta}$, where $\beta$ is an element of GF$(q^2)$ such that $(\beta^q - \beta)^2$ is a nonsquare in GF$(q)$. In particular, $\beta \notin$ GF$(q)$ and the result follows from Theorem 4.19. $\square$

It should be noted that orthogonal-Buekenhout-Metz unitals which can be expressed as a union of conics were independently discovered by Baker and Ebert in [21] and by Hirschfeld and Szőnyi in [133].

### 4.2.2 The Even Characteristic Case

We turn our attention to ovoidal-Buekenhout-Metz unitals embedded in PG$(2, q^2)$, where $q = 2^e$ is an even prime power. We follow Ebert [108] for orthogonal-Buekenhout-Metz unitals and Ebert [109] for Buekenhout-Tits unitals. First consider the orthogonal case, where the ovoidal cone in PG$(4, q)$ is an elliptic cone. Here the situation is almost identical to that when $q$ is an odd prime power, except that one replaces the notion of quadratic character (squares and nonsquares) by the notion of absolute trace, as defined in Section 1.2. Of course, we also need to choose a different basis for the vector space GF$(q^2)$ over its subfield GF$(q)$. To establish some notation, we let $T_0$ and $T_1$ denote the elements in GF$(q) =$ GF$(2^e)$ that have absolute trace 0 and 1, respectively. Thus GF$(q)$ is the disjoint union of $T_0$ and $T_1$. When computing in characteristic 2, we must remember that $2 = 0$ and thus $x + y = x - y$ for all field elements $x$ and $y$.

**Lemma 4.21.** *Assume that $q \geq 4$ is a power of 2. Then there is some element $\delta \in$ GF$(q^2) \backslash$ GF$(q)$ such that $\delta^q = 1 + \delta$ and $\delta^2 = v + \delta$ for some $v \in T_1$ with $v \neq 1$.*

*Proof.* Using the trace function from GF$(q^2)$ to GF$(q)$ as discussed in Section 1.2, there are $q$ solutions in GF$(q^2)$ of the equation $x^q + x + 1 = 0$. Since $q \geq 4$, we may choose a solution $\delta$ of this equation whose multiplicative order is not 3. In particular, $\delta^q = 1 + \delta$ and hence $\delta \notin$ GF$(q)$. Moreover, $\delta^2 + \delta = (\delta^q + 1)^2 + (\delta^q + 1) = (\delta^2 + \delta)^q$ and thus $\delta^2 + \delta$ is some element $v \in$ GF$(q)$. Since $\delta$ is a solution of the quadratic equation $x^2 + x + v = 0$ and $\delta \notin$ GF$(q)$, this quadratic equation must be irreducible over GF$(q)$. Therefore $v \in T_1$, as discussed in Section 1.2. Finally, if $v = 1$, then $\delta$ is a solution of $x^2 + x + 1 = 0$ and hence $v$ has multiplicative order 3, contradicting our choice of $\delta$. This completes the proof. $\square$

We now take $\{1, \delta\}$ as our basis for GF$(q^2)$ over GF$(q)$. As in the previous section, for each $\alpha, \beta \in$ GF$(q^2)$, we define

$$U_{\alpha\beta} = \{(x, \alpha x^2 + \beta x^{q+1} + r, 1) \mid x \in \text{GF}(q^2), r \in \text{GF}(q)\} \cup \{0, 1, 0)\}. \quad (4.4)$$

Using the above basis for $GF(q^2)$ over $GF(q)$, we expand using Lemma 4.21 and obtain the following equation for $U_{\alpha\beta}$:

$$(a_1 + b_1)x_0^2 + b_1 x_0 x_1 + (a_0 + a_1 + a_1 v + b_1 v)x_1^2 + y_1 = 0. \qquad (4.5)$$

Thinking of $(x_0, x_1, y_0, y_1, 1)$ as the affine point of $PG(4, q) \backslash \Sigma_\infty$ corresponding to the affine point $(x, y, 1)$ of $PG(2, q^2) \backslash \ell_\infty$ in the Bruck-Bose correspondence of Section 3.4.4, the above equation is once again of the desired form $f(X_0, X_1) + X_3 X_4 = 0$, where $f$ is some binary quadratic form and $X_0 = x_0$, $X_1 = x_1$, $X_3 = y_1$, $X_4 = 1$.

Moreover, as discussed in Section 1.2, $f$ is irreducible if and only if $b_1 \neq 0$ (that is, $\beta \notin GF(q)$) and $(a_1 + b_1)(a_0 + a_1 + a_1 v + b_1 v)/b_1^2 \in T_1$. Using the additive property of the absolute trace $T$ and the fact that

$$\frac{a_0 b_1 + a_1 b_1 + a_0^2 + a_1^2}{b_1^2} = \frac{a_0 + a_1}{b_1} + \left(\frac{a_0 + a_1}{b_1}\right)^2$$

has absolute trace 0, this condition is equivalent to

$$T\left(\frac{\alpha^{q+1} + (\beta^q + \beta)^2 v}{(\beta^q + \beta)^2}\right) = T\left(\frac{a_0^2 + a_0 a_1 + (a_1^2 + b_1^2)v}{b_1^2}\right) = 1.$$

Since $T(v) = 1$, the latter condition is in turn equivalent to

$$T\left(\frac{\alpha^{q+1}}{(\beta + \beta)^2}\right) = 0.$$

Analogously to the odd characteristic case, in this section we let $d$ denote the element $\alpha^{q+1}/(\beta^q + \beta)^2$ in $GF(q)$, whenever $\beta \notin GF(q)$, and think of $d$ as the *discriminant* of $U_{\alpha\beta}$ when $q$ is even. We have thus shown that whenever $\beta \notin GF(q)$ and $d \in T_0$, the set of points $U_{\alpha\beta}$ as defined in (4.4) is an orthogonal-Buekenhout-Metz unital embedded in $PG(2, q^2)$ which is tangent to $\ell_\infty = (0, 0, 1)'$ at the point $P_\infty = (0, 1, 0)$. Conversely, using equation (4.5), we can solve for an appropriate $\alpha$ and $\beta$, given any irreducible binary quadratic form $f$. We thus have the following result.

**Theorem 4.22.** *Let $q \geq 4$ be a power of 2. Then every orthogonal-Buekenhout-Metz unital in $PG(2, q^2)$ is equivalent to one expressed as $U_{\alpha\beta}$ for some $\alpha, \beta$ in $GF(q^2)$, where $\beta \notin GF(q)$ and $d = \alpha^{q+1}/(\beta^q + \beta)^2 \in T_0$. Conversely, every such $U_{\alpha\beta}$ is an orthogonal-Buekenhout-Metz unital in $PG(2, q^2)$.*

Just as for odd $q$, $U_{\alpha\beta}$ is classical if (and only if) $\alpha = 0$. Moreover, the stabilizer of $U_{\alpha\beta}$ in $PGL(3, q^2)$ (or $P\Gamma L(3, q^2)$) is computed in a completely analogous fashion. We state the following theorem without proof (details can be found in [108]).

**Theorem 4.23.** *Let $q = 2^e$ for some integer $e \geq 2$, and let $U_{\alpha\beta}$ be an orthogonal-Buekenhout-Metz unital in PG$(2, q^2)$ as defined by (4.4). Let $F$ denote GF(2) if $\alpha = 0$ or the smallest subfield of GF$(q)$ containing the element $d = \alpha^{q+1}/(\beta^q + \beta)^2$ if $\alpha \neq 0$. Let $G$ be the subgroup of P$\Gamma$L$(3, q^2)$ fixing $P_\infty$ and leaving $U_{\alpha\beta}$ invariant. Then the following hold.*

1. *The order of $G$ is $mq^3(q^2 - 1)$ or $mq^3(q - 1)$, according to $\alpha = 0$ or $\alpha \neq 0$, where $m$ is the dimension of GF$(q^2)$ over its subfield $F$.*
2. *The group $G$ has point orbits $U_{\alpha\beta} \backslash P_\infty$, $\ell_\infty \backslash P_\infty$, and PG$(2, q^2) \backslash (U_{\alpha\beta} \cup \ell_\infty)$.*
3. *If $\alpha \neq 0$ (hence $U_{\alpha\beta}$ is nonclassical), then every collineation of PG$(2, q^2)$ which stabilizes $U_{\alpha\beta}$ necessarily fixes $P_\infty$; that is, $G$ is the full stabilizer of $U_{\alpha\beta}$ if $\alpha \neq 0$.*
4. *If $G_0 = G \cap$ PGL$(3, q^2)$ is the homography subgroup of $G$, then the index of $G_0$ in $G$ is $m$. More precisely, $G_0$ is the semidirect product of a non-abelian subgroup $S$ of order $q^3$ by a cyclic subgroup $J$ of order $q^2 - 1$ or $q - 1$ according to $\alpha = 0$ or $\alpha \neq 0$.*

It should be noted that the cyclic subgroup $J$ of Theorem 4.23 is generated by the homography

$$(x, y, z) \mapsto (\eta x, \ \eta^{q+1} y, \ z),$$

where $\eta = \zeta$ is a primitive element of GF$(q^2)$ if $\alpha = 0$, or $\eta = w$ is a primitive element of GF$(q)$ if $\alpha \neq 0$. Moreover, the subgroup $S$ is always non-abelian because $\beta$ is never an element of GF$(q)$ when $q$ is even. In fact, one can say a bit more about the Sylow 2-subgroup $S$ of $G_0$ in the even characteristic case.

**Corollary 4.24.** *For even $q \geq 4$, the Sylow 2-subgroup $S$ of $G_0$ is special.*

*Proof.* The matrices inducing the elements of $S$ can be computed exactly as for odd $q$, simply replacing 2 by 0 and $-$ by $+$ in all the entries of the matrices. Thus $S$ an elementary abelian 2-subgroup of order $q$, namely, the elation subgroup $K$ with axis $\ell_\infty$ and center $P_\infty$, which turns out to be the center of $S$ when $q$ is even. One can then compute directly that the $q^3 - q$ elements in $S \backslash K$ all have order 4, and the only elements of order 2 in $S$ are the nonidentity elements of $K$. From this it follows that $S$ is special.   □

The equivalences among the unitals $U_{\alpha\beta}$ are sorted out in the even characteristic case exactly as they were in the odd characteristic case. In fact, one has the same equation (4.3) for equivalence. The following theorem is stated here without proof. Details of the argument, if needed, can be found in [108].

**Theorem 4.25.** *Let $q = 2^e$ for some integer $e \geq 2$. Then the number of mutually inequivalent orthogonal-Buekenhout-Metz unitals of order $q$ in PG$(2, q^2)$ is*

$$\frac{1}{2e} \sum_{m|e} \varphi\left(\frac{2e}{m}\right) 2^m,$$

*where $\varphi$ is the Euler totient function.*

**Corollary 4.26.** *If $q = 2^e$ for some prime e, then the number of mutually inequivalent orthogonal-Buekenhout-Metz unitals of order q in $\mathrm{PG}(2,q^2)$ is 2 if $e = 2$ and is $1 + (2^{e-1} - 1)/e$ if e is an odd prime.*

We now study the tangent lines to $U_{\alpha\beta}$ in the even characteristic case.

**Lemma 4.27.** *Let $q \geq 4$ be a power of 2. Then the line in $\mathrm{PG}(2,q^2)$ tangent to $U_{\alpha\beta}$ at the point $(x, \alpha x^2 + \beta x^{q+1} + r, 1)$ is $((\beta^q + \beta)x^q, 1, \alpha x^2 + \beta^q x^{q+1} + r)'$.*

*Proof.* As in the proof of Lemma 4.16, we first prove that $\ell = (0,1,0)'$ is the tangent line to $U_{\alpha\beta}$ at the point $P = (0,0,1)$. Clearly $P \in \ell$, and any other point of $\ell \cap U_{\alpha\beta}$ looks like $(x,0,1)$ for some nonzero $x \in \mathrm{GF}(q^2)$. But then (4.4) implies that $\alpha x^2 + \beta x^{q+1} \in \mathrm{GF}(q)$ and hence

$$\alpha x^2 + \beta x^{q+1} = \alpha^q x^{2q} + \beta^q x^{q+1}.$$

This further implies that $y^2 + y = d$, where $d = \alpha^{q+1}/(\beta^q + \beta)^2$ is the discriminant of $U_{\alpha\beta}$ and $y = \alpha/(\beta^q + \beta)x^{q-1}$. Since $y$ is a solution of the quadratic equation $Y^2 + Y + d = 0$ over $\mathrm{GF}(q)$ and since $d \in T_0$, we know from Section 1.2 that this equation factors over $\mathrm{GF}(q)$ and hence $y \in \mathrm{GF}(q)$. This further implies that $\alpha/x^{q-1} \in \mathrm{GF}(q)$ and therefore $\alpha^q x^{2q} = \alpha x^2$. Hence $(\beta^q + \beta)x^{q+1} = \alpha x^2 + \alpha^q x^{2q} = 0$, implying $x = 0$ as $\beta \notin \mathrm{GF}(q)$. This contradicts our choice of $x$, and therefore $(0,1,0)'$ is the unique tangent line to $U_{\alpha\beta}$ at the point $(0,0,1)$.

The result now follows exactly as in the proof of Lemma 4.16. In fact, the dual coordinates for the tangent line at $(x, \alpha x^2 + \beta x^{q+1} + r, 1)$ are exactly the same as for odd $q$, if the coordinates are read in a field of characteristic 2.    □

**Theorem 4.28.** *Let $q \geq 4$ be a power of 2. Then the orthogonal-Buekenhout-Metz unital $U_{\alpha\beta}$ in $\mathrm{PG}(2,q^2)$ defined by (4.4) is self-dual.*

*Proof.* The proof is exactly the same as for Theorem 4.17. One only has to interpret all computations in a field of characteristic 2.    □

**Theorem 4.29.** *Let $q \geq 4$ be a power of 2, and let $U_{\alpha\beta}$ be an orthogonal-Buekenhout-Metz unital of $\mathrm{PG}(2,q^2)$ which is not classical. Let $Q \in \mathrm{PG}(2,q^2) \backslash U_{\alpha\beta}$. Then the feet of Q are collinear if and only if $Q \in \ell_\infty$.*

*Proof.* The proof is completely analogous to that of Theorem 4.18.    □

For odd $q$, there is one class of orthogonal-Buekenhout-Metz unitals of order $q$ that can be expressed as a union of conics in $\mathrm{PG}(2,q^2)$, pairwise meeting at the point $P_\infty$. This does not happen for even $q$. Rather than considering only conics, we consider the more general problem of determining if an oval, as defined in Section 1.4, can be embedded in some $U_{\alpha\beta}$. In Section 1.3 we discussed that for even $q$, the tangents to a conic in a Desarguesian projective plane all pass through a common point, the nucleus of the conic. The same holds true for the tangents to an oval in even characteristic (see [131]). Of course, the combinatorics dictate that there is a unique tangent line at each point of an oval, just as for conics.

**Theorem 4.30.** *Let $q \geq 4$ be a power of 2, and let $U_{\alpha\beta}$ be an orthogonal-Buekenhout-Metz unital in PG$(2, q^2)$. Then $U_{\alpha\beta}$ contains no oval of PG$(2, q^2)$. In particular, the Hermitian curve in PG$(2, q^2)$, for even $q \geq 4$, contains no irreducible conic.*

*Proof.* Suppose $U_{\alpha\beta}$ contains an oval $\mathcal{O}$. Since $q$ is even, the $q^2 + 1$ tangents to $\mathcal{O}$ are concurrent at some point of PG$(2, q^2)$, namely, the nucleus $N$. Since the unique tangent line to $U_{\alpha\beta}$ at each point of $\mathcal{O}$ is also the unique tangent to $\mathcal{O}$ at that point, we have $q^2 + 1$ tangent lines to $U_{\alpha\beta}$ passing through some point of PG$(2, q^2)$. This contradicts the fact that every point of PG$(2, q^2)$ lies on exactly 1 or $q + 1$ tangents to the embedded unital $U_{\alpha\beta}$.     □

We now assume that $q = 2^e$, where $e \geq 3$ is an odd integer. Let $\tau$ be the automorphism of GF$(q)$ defined by

$$\tau : x \rightarrow x^{2^{(e+1)/2}}.$$

Thus $\tau^2 : x \rightarrow x^2$ for all $x \in$ GF$(q)$. Then the Suzuki-Tits ovoid [212] in PG$(3, q)$, as introduced in Section 1.4, may be coordinatized as

$$O_T = \{(0, 0, 1, 0)\} \cup \left\{ (x_0, x_1, x_0^{\tau+2} + x_1^{\tau} + x_0 x_1, 1) : x_0, x_1 \in GF(q) \right\}.$$

As shown in [212], every nontrivial planar section of $O_T$ is a nonconic oval, and the unique tangent plane to $O_T$ at the point $(x_0, x_1, x_0^{\tau+2} + x_1^{\tau} + x_0 x_1, 1)$ is

$$(x_1, x_0, 1, x_0^{\tau+2} + x_1^{\tau} + x_0 x_1)'.$$

The tangent plane to $O_T$ at $(0, 0, 1, 0)$ is $(0, 0, 0, 1)'$.

Using the Bruck-Bose representation for PG$(2, q^2)$ in $\Sigma = $ PG$(4, q)$, as coordinatized in Section 3.4.4, we embed $O_T$ in $\Sigma$ by taking $(0, 0, 1, 0, 0)'$ as the hyperplane containing $O_T$. We now construct a cone in $\Sigma$ whose base is this embedded Tits ovoid and whose vertex is the point $V = (0, 0, 1, 0, 0)$. The resulting ovoidal cone is

$$C = \{(0, 0, 0, 1, 0)\} \cup \{(0, 0, 1, y_1, 0) \mid y_1 \in GF(q)\}$$
$$\cup \left\{ (x_0, x_1, y_0, x_0^{\tau+2} + x_1^{\tau} + x_0 x_1, 1) \mid x_0, x_1, y_0 \in GF(q) \right\}.$$

The unique tangent hyperplane through the line

$$p_\infty = \langle (0, 0, 1, 0, 0), (0, 0, 0, 1, 0) \rangle$$

of $C$ is $\Sigma_\infty$, whose equation is $X_4 = 0$. Since the regular spread $\mathcal{S}$ of $\Sigma_\infty$, used in the Bruck-Bose representation for PG$(2, q^2)$, contains the generator $p_\infty$, we see from Section 4.1 that $C$ represents an ovoidal-Buekenhout-Metz unital $U$ in $PG(2, q^2)$. In fact, this is a Buekenhout-Tits unital, as defined in Section 4.1, and we denote it by $U_T$. Using the same basis $\{1, \delta\}$ for $GF(q^2)$ over $GF(q)$, as given by Lemma 4.21, we can express $U_T$ as

$$U_T = \left\{ (x_0 + x_1\delta,\ y_0 + (x_0^{\tau+2} + x_1^\tau + x_0x_1)\delta,\ 1) \mid x_0, x_1, y_0 \in GF(q) \right\}$$
$$\cup \{(0, 1, 0)\}. \tag{4.6}$$

We now compute the stabilizer of $U_T$ in $PGL(3, q^2)$, which we shall see is considerably smaller than the homography stabilizer of the orthogonal-Buekenhout-Metz unitals. For each $a, b \in GF(q)$, let $\vartheta_{a,b}$ denote the homography of $PG(2, q^2)$ given by

$$(x, y, z) \mapsto (x + b\delta z,\ b(1 + \delta)x + y + (a + b\delta)z,\ z).$$

Straightforward computations show that $\vartheta_{a,b}\vartheta_{c,d} = \vartheta_{a+c+vbd,\ b+d}$, where $v = \delta^2 + \delta$ as in Lemma 4.21. Hence $\vartheta_{a,b}^{-1} = \vartheta_{a+vb^2,\ b}$. In particular, $G_T = \{\vartheta_{a,b} \mid a, b \in GF(q)\}$ is an abelian group of order $q^2$.

**Theorem 4.31.** *The abelian group $G_T$ of order $q^2$ stabilizes the Buekenhout-Tits unital $U_T$ in $PG(2, q^2)$. In particular, the element $\vartheta_{a,b} \in G_T$ fixes the point $P_\infty = (0, 1, 0)$ and maps the point $P_{s,t,r} = (s + t\delta,\ r + (s^{\tau+2} + t^\tau + st)\delta,\ 1) \in U_T$ to the point $P_{s,t+b,r+a+bs+vbt} \in U_T$, for each choice of $r, s, t \in GF(q)$. Moreover, $G_T$ has $q$ point orbits of size $q^2$ on $U_T \setminus P_\infty$, has $q$ orbits of size $q$ on $\ell_\infty \setminus P_\infty$, and has $q^2 - q$ orbits of size $q^2$ on $PG(2, q^2) \setminus (U_T \cup \ell_\infty)$.*

*Proof.* The action of $G_T$ on the points of $U_T$ follows from a straightforward computation. Since $G_T$ fixes the point $P_\infty \in U_T$, it must also fix the unique tangent line to $U_T$ at $P_\infty$, namely, $\ell_\infty$. Another elementary computation shows that $G_T$ acts fixed-point-freely on $PG(2, q^2) \setminus \ell_\infty$. On the other hand, if $(1, y, 0) \in (\ell_\infty \setminus P_\infty)$, the $G_T$-stabilizer of this point is $\{\vartheta_{a,0} \mid a \in GF(q)\}$. The result now follows from the Orbit-Stabilizer Theorem. $\square$

We now consider the tangent lines to the unital $U_T$ and the associated question of self-duality. Of course, from the ovoidal Buekenhout construction we know that $\ell_\infty = (0, 0, 1)'$ is the unique tangent line to $U_T$ at the point $P_\infty = (0, 1, 0)$.

**Lemma 4.32.** *For the Buekenhout-Tits unital $U_T$, the unique tangent line at the point $(x_0 + x_1\delta,\ y_0 + (x_0^{\tau+2} + x_1^\tau + x_0x_1)\delta,\ 1)$ is*

$$(x_0 + x_1 + x_1\delta,\ 1,\ x_0^2 + x_1^2v + x_0x_1 + y_0 + (x_0^{\tau+2} + x_1^\tau + x_0x_1)\delta)',$$

*where $v = \delta^2 + \delta$ as in Lemma 4.21.*

*Proof.* Let $P = (x_0, x_0^{\tau+2}\delta, 1)$ for some $x_0 \in$ GF$(q)$. Then $P \in U_T$ by equation (4.6). We claim that the unique tangent line to $U_T$ at $P$ is $\ell = (x_0, 1, x_0^2 + x_0^{\tau+2}\delta)'$. It is clear that $P \in \ell$, and we suppose that $P_{s,t,r} = (s + t\delta, r + (s^{\tau+2} + t^\tau + st)\delta, 1)$ is another point of $U_T \cap \ell$, where $r, s, t \in$ GF$(q)$. Then necessarily

$$x_0(s + t\delta) + r + (s^{\tau+2} + t^\tau + st)\delta + x_0^2 + x_0^{\tau+2}\delta = 0,$$

which further implies that

(i) $x_0 s + x_0^2 + r = 0$,
(ii) $x_0 t + s^{\tau+2} + t^\tau + st + x_0^{\tau+2} = 0$.

Returning to the Tits ovoid $O_T$, we see that $(x_0, 0, x_0^{\tau+2}, 1) \in O_T$ and the unique plane in PG$(3, q)$ tangent to $O_T$ at this point is $(0, x_0, 1, x_0^{\tau+2})'$. Since the point $(s, t, s^{\tau+2} + t^\tau + st, 1) \in O_T$ also lies on this plane by equation (ii) above, we must have $s = x_0$ and $t = 0$, which further implies by equation (i) that $r = 0$. Hence $P_{s,t,r} = P$ and $\ell$ is the unique tangent line to $U_T$ at $P$.

Now consider the element $\vartheta_{y_0+x_0x_1, x_1} \in G_T$ from Theorem 4.31. Then $\vartheta_{y_0+x_0x_1, x_1}$ stabilizes $U_T$, maps the point $P = P_{x_0,0,0}$ to the point $P_{x_0,x_1,y_0}$, and hence maps the line $\ell = (x_0, 1, x_0^2 + x_0^{\tau+2}\delta)'$ to the tangent line at the point

$$(x_0 + x_1\delta, \ y_0 + (x_0^{\tau+2} + x_1^\tau + x_0x_1)\delta, \ 1).$$

Using the fact that $\vartheta_{y_0+x_0x_1, x_1}^{-1} = \vartheta_{y_0+x_0x_1+vx_1^2, x_1}$, the result now follows from a direct computation.    □

**Theorem 4.33.** *Let $U_T$ be a Buekenhout-Tits unital in PG$(2, q^2)$, and let $R$ be any point of PG$(2, q^2)\backslash U_T$. Then the feet of $R$ are collinear if and only if $R$ lies on $\ell_\infty$.*

*Proof.* The proof is very similar to that given for the analogous result concerning orthogonal-Buekenhout-Metz unitals. From Lemma 4.32, we know that the tangents to $U_T$, other than $\ell_\infty$, are the $q^3$ lines with dual coordinates

$$\ell_{s,t,r} = (s + t + t\delta, \ 1, \ r + s^2 + vt^2 + st + (s^{\tau+2} + t^\tau + st)\delta)',$$

for $r, s, t \in$ GF$(q)$. Adopting the notation used in the proof of Lemma 4.32, the foot of the tangent line $\ell_{s,t,r}$ is $P_{s,t,r}$.

Suppose first that $R \in \ell_\infty$. Since $R \notin U_T$, we must have $R = (1, y, 0)$ for some $y \in$ GF$(q^2)$. Using the basis $\{1, \delta\}$ for GF$(q^2)$ over GF$(q)$, as given by Lemma 4.21, we express $y$ uniquely as $y_0 + y_1\delta$ for some $y_0, y_1 \in$ GF$(q)$. Then the tangent lines through $R$, other than $\ell_\infty$, are the $q$ lines $\ell_{s,t,r}$ whose parameters satisfy the condition $s + t + y_0 + (t + y_1)\delta = 0$, implying that $t = y_1$ and $s = y_0 + y_1$. Therefore the feet of $R$ are

$$\{P_{y_0+y_1, y_1, r} \mid r \in \text{GF}(q)\} \cup \{P_\infty\}.$$

These feet clearly lie on the line joining $P_\infty$ to $P_{y_0+y_1, y_1, 0}$, and hence are collinear.

Conversely, suppose $R \notin \ell_\infty$. Then $R = (x, y, 1)$ for some $x, y \in \mathrm{GF}(q^2)$. Expressing $x = x_0 + x_1 \delta$ and $y = y_0 + y_1 \delta$ uniquely for $x_0, x_1, y_0, y_1 \in \mathrm{GF}(q)$, the $q + 1$ tangent lines incident with $R$ are easily seen to be the lines $\ell_{s,t,r}$, where $r, s, t \in \mathrm{GF}(q)$ satisfy

$$s^2 + st + vt^2 + x_0 s + (x_0 + vx_1)t + y_0 + r = 0,$$
$$s^{\tau+2} + t^\tau + st + x_1 s + x_0 t + y_1 = 0.$$

Solving for $r$ in the first equation, the corresponding $q + 1$ feet are

$$\mathcal{F} = \{(s + t\delta, \ s^2 + st + vt^2 + x_0 s + x_0 t + vx_1 t + y_0 + (s^{\tau+2} + t^\tau + st)\delta, \ 1) \\ \mid s^{\tau+2} + t^\tau + st = x_1 s + x_0 t + y_1\}.$$

If these feet were incident with a line of the form $(\alpha, 1, \beta)'$, for some $\alpha, \beta \in \mathrm{GF}(q^2)$, then by expressing $\alpha = a_0 + a_1 \delta$ and $\beta = b_0 + b_1 \delta$ for $a_0, a_1, b_0, b_1 \in \mathrm{GF}(q)$, we obtain

(i) $s^2 + st + vt^2 + (a_0 + x_0)s + (x_0 + vx_1 + va_1)t + y_0 + b_0 = 0,$
(ii) $(a_1 + x_1)s + (a_0 + a_1 + x_0)t + y_1 + b_1 = 0.$

Viewing the ordered pair $(s, t)$ as a point in the classical affine plane $\mathrm{AG}(2, q)$ of order $q$, equation (i) represents the $q + 1$ points $(s, t)$ on some ellipse since $v \in \mathcal{T}_1$ and thus $s^2 + st + vt^2$ is an irreducible binary quadratic form. On the other hand, equation (ii) represents the points $(s, t)$ on some line. Since the $q + 1$ ordered pairs $(s, t)$ corresponding to the feet $F$ must satisfy both (i) and (ii), we arrive at an obvious contradiction.

Similarly, if the feet $F$ lie on a line of the form $(1, 0, \alpha)$, the corresponding ordered pairs $(s, t)$ satisfy the equation $s + t\delta + a_0 + a_1\delta = 0$, implying that $s = a_0$ and $t = a_1$. This again contradicts the fact that we must have $q + 1$ choices for $(s, t)$. Therefore, all cases considered, the feet of a point $R \notin \ell_\infty$ do not form a collinear set. $\qquad\square$

**Corollary 4.34.** *The group $G_T$ of Theorem 4.31 is the full homography stabilizer of the Buekenhout-Tits unital $U_T$ in $\mathrm{PG}(2, q^2)$.*

*Proof.* From Theorem 4.31, we know that the homography subgroup $G_T$ leaves $U_T$ invariant. Conversely, by Theorem 4.33, we know that any homography stabilizing $U_T$ must leave $\ell_\infty$ invariant and hence fix the point $P_\infty = U_T \cap \ell_\infty$. Therefore, any homography stabilizing $U_T$ must be of the form

$$(x, y, z): \ \longmapsto (\theta_1 x + \theta_2 z, \ \theta_3 x + \theta_4 y + \theta_5 z, \ z),$$

where $\theta_1, \theta_2, \theta_3, \theta_4, \theta_5 \in \mathrm{GF}(q^2)$ with $\theta_1 \theta_4 \neq 0$. Careful consideration of the conditions imposed on these parameters by the stabilization of $U_T$ enables one to show that any such homography in the stabilizer must indeed be an element of $G_T$ (see [109] for details). $\qquad\square$

**Corollary 4.35.** *The Buekenhout-Tits unital $U_T$ in PG$(2, q^2)$ is self-dual.*

*Proof.* The points of $U_T^*$ are the tangent lines of $U_T$, namely, $\{(x_0 + x_1 + x_1\delta,\ 1,\ x_0^2 + x_1^2 v + x_0 x_1 + y_0 + (x_0^{\tau+2} + x_1^\tau + x_0 x_1)\delta)' \mid x_0, x_1, y_0 \in \mathrm{GF}(q)\} \cup \{(0,0,1)'\} = \{(x_0 + x_1 + x_1\delta,\ 1,\ y_0 + (x_0^{\tau+2} + x_1^\tau + x_0 x_1)\delta)' \mid x_0, x_1, y_0 \in \mathrm{GF}(q)\} \cup \{(0,0,1)'\}$. On the other hand, the image of the unital $U_T$ under the semilinear collineation induced by the conjugation field automorphism $x \mapsto x^q$ followed by interchanging second and third coordinates is $\{(x_0 + x_1 + x_1\delta,\ 1,\ y_0 + x_0^{\tau+2} + x_1^\tau + x_0 x_1 + (x_0^{\tau+2} + x_1^\tau + x_0 x_1)\delta) \mid x_0, x_1, y_0 \in \mathrm{GF}(q)\} \cup \{(0,0,1)\} = \{(x_0 + x_1 + x_1\delta,\ 1,\ y_0 + (x^{\tau+2} + x_1^\tau + x_0 x_1)\delta) \mid x_0, x_1, y_0 \in \mathrm{GF}(q)\} \cup \{(0,1,0)\}$, since $\delta^q = 1 + \delta$ by Lemma 4.21. Hence $U_T$ and $U_T^*$ are isomorphic as unitals, proving the result. $\square$

We conclude this section by observing that Buekenhout-Tits unitals contain no ovals, just as was true for even order orthogonal-Buekenhout-Metz unitals.

**Theorem 4.36.** *The Buekenhout-Tits unital $U_T$ in PG$(2, q^2)$ contains no oval of PG$(2, q^2)$.*

*Proof.* The proof follows exactly as in the proof of Theorem 4.30. $\square$

We close this chapter with a discussion in a general setting concerning a configuration first found by O'Nan. An **O'Nan configuration** is a collection of four distinct lines meeting in six distinct points, as illustrated in Figure 4.2.

**Fig. 4.2.** An O'Nan configuration

Such configurations first appeared in [174], where O'Nan observed that the classical unital contains no such configuration. In fact, it is now conjectured that this property characterizes classical unitals.

It is known that if $U$ is an ovoidal-Buekenhout-Metz unital with respect to the point $P_\infty = U \cap \ell_\infty$, then $U$ contains no O'Nan configurations through $P_\infty$ (a simple proof of this using the Bruck-Bose representation is given in the proof of Lemma 7.42 later in the book). There are no known constructions of O'Nan configurations in an arbitrary nonclassical ovoidal-Buekenhout-Metz unital embedded in PG$(2, q^2)$, although computationally it is known that such configurations do exist for small values of $q$.

# 5

# Unitals Embedded in Non-Desarguesian Planes

In this chapter we discuss some known constructions of unitals embedded in non-Desarguesian projective planes of square order, some of which are translation planes and some of which are not. As discussed in Section 4.1, ovoidal-Buekenhout-Metz unitals exist in any two-dimensional translation plane, and nonsingular-Buekenhout unitals exist in certain derivable two-dimensional translation planes. Here we discuss a wide variety of constructions, not just the Buekenhout techniques. We also show that it is possible for a Buekenhout unital to be embedded in two nonisomorphic planes.

## 5.1 Unitals in Hall Planes

The Hall plane of order $q^2$ can be obtained by deriving the Desarguesian plane $PG(2, q^2)$ of order $q^2$ as discussed in Section 3.2. We review the notation here. Let $D$ be a derivation set of $\ell_\infty$ in $PG(2, q^2)$. That is, $D$ is a Baer subline of $\ell_\infty \cong PG(1, q^2)$. The Hall plane, $\text{Hall}(q^2)$, has as *points* the points of $PG(2, q^2) \setminus D$ and $q + 1$ new points denoted by $D'$. The *lines* of $\text{Hall}(q^2)$ are of three types: lines of $PG(2, q^2)$ that meet $\ell_\infty$ in a point not in $D$; Baer subplanes of $PG(2, q^2)$ that contain $D$; the line at infinity, $\ell'_\infty$, which consists of the points of $\ell_\infty \setminus D$ and the points in $D'$ (the latter points correspond to the new slope points). *Incidence* in $\text{Hall}(q^2)$ is inherited from $PG(2, q^2)$. Of course, we may choose $\ell_\infty$ to be any line of $PG(2, q^2)$.

Thus, as shown in Section 3.4.3, we may assume the following Bruck-Bose representation for the Hall plane. Let $S$ be a regular spread in the hyperplane $\Sigma_\infty$ of $PG(4, q)$, so that $\mathcal{P}(S)$ is the Desarguesian plane $PG(2, q^2)$. Let $\mathcal{R}$ be a regulus of $S$ with opposite regulus $\mathcal{R}'$. Then $S' = (S \setminus \mathcal{R}) \cup \mathcal{R}'$ is a spread of $\Sigma_\infty$ that constructs the Hall plane; that is, $\mathcal{P}(S') \cong \text{Hall}(q^2)$. Hence the Hall plane is a translation plane of dimension two over its kernel. This plane is isomorphic to the one coordinatized by the Hall quasifield. See [140] for more details on the Hall plane.

S. Barwick, G. Ebert, *Unitals in Projective Planes*,
DOI: 10.1007/978-0-387-76366-8_5, © Springer Science+Business Media, LLC 2008

Recall that the only known unitals embedded in the Desarguesian plane are Buekenhout unitals, and the only nonsingular-Buekenhout unital embedded in the Desarguesian plane is the classical unital. If $U$ is any unital embedded in $PG(2, q^2)$, then when we derive this Desarguesian plane with respect to a derivation set $D$ on $\ell_\infty$, we obtain a corresponding set of points $U'$ in $\text{Hall}(q^2)$. To define $U'$ more precisely, let $\text{aff}U$ be the affine points of $U$; that is, $\text{aff}U = U \setminus \ell_\infty$ (that is, $U$ minus the points of $U$ that lie on $\ell_\infty$). Since the affine points of $PG(2, q^2)$ are the same as the affine points of $\text{Hall}(q^2)$, we have $\text{aff}U' = \text{aff}U$ in $\text{Hall}(q^2)$. We complete $\text{aff}U'$ to a set $U'$ in $\text{Hall}(q^2)$ by adding the points on the line at infinity $\ell'_\infty$ of $\text{Hall}(q^2)$ which lie on a $q$-secant of $\text{aff}U'$. We want to determine when the set $U'$ is a unital in $\text{Hall}(q^2)$.

We begin by focusing on the classical unital embedded in $PG(2, q^2)$. Every chord of a classical unital is a Baer subline and so can serve as a derivation set. Hence there are five different ways to position the derivation set in relation to a classical unital $U$ embedded in $PG(2, q^2)$; these cases are illustrated in Figure 5.1.

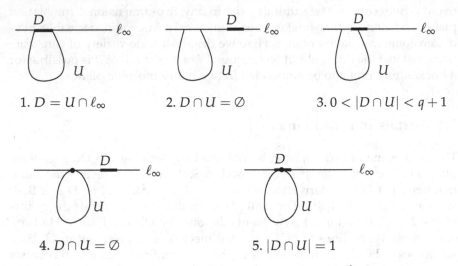

1. $D = U \cap \ell_\infty$    2. $D \cap U = \emptyset$    3. $0 < |D \cap U| < q + 1$

4. $D \cap U = \emptyset$    5. $|D \cap U| = 1$

**Fig. 5.1.** Deriving the classical unital in $PG(2, q^2)$

We first consider the case when $U$ is secant to $\ell_\infty$. Then $U$ corresponds to a nonsingular quadric $\mathcal{U}$ in $PG(4, q)$ that meets $\Sigma_\infty$ in a hyperbolic quadric which is ruled by a regulus $\mathcal{R}$ of the regular spread $\mathcal{S}$. Of course, the opposite regulus $\mathcal{R}'$ also rules this hyperbolic quadric. Now $D = U \cap \ell_\infty$ is a Baer subline, and thus is a derivation set. If we derive $PG(2, q^2)$ with respect to $D$, then we obtain the Hall plane which is constructed in the Bruck-Bose representation by the spread $\mathcal{S}' = (\mathcal{S} \setminus \mathcal{R}) \cup \mathcal{R}'$. The points of $U'$ in $\text{Hall}(q^2)$ correspond to the same nonsingular quadric $\mathcal{U}$, which "meets" the spread $\mathcal{S}'$

in the regulus $\mathcal{R}'$. Hence $U'$ is a nonsingular-Buekenhout unital of Hall($q^2$) by Theorem 4.2. Grüning [127] studied this unital in detail.

**Theorem 5.1.** *Let $U$ be a classical unital in* PG($2, q^2$) *secant to $\ell_\infty$. If we derive* PG($2, q^2$) *with respect to $U \cap \ell_\infty$, then $U'$ is a nonsingular-Buekenhout unital in* Hall($q^2$). *Furthermore, $U'$ contains no O'Nan configurations with 0, 2, or 3 points on $\ell'_\infty$, but $U'$ does contain O'Nan configurations with 1 point on $\ell'_\infty$.*

See [127] for the proof of the existence of O'Nan configurations. In particular, it is shown that if $P$ is a point of $U' \cap \ell'_\infty$, if $m_1$ and $m_2$ are two chords of $U$ through $P$ (distinct from $\ell'_\infty$), and if $m_3$ is a chord that meets $m_1$ and $m_2$ but not $U' \cap \ell'_\infty$, then there is an O'Nan configuration in $U'$ containing $m_1, m_2, m_3$. Hence, as the classical unital contains no O'Nan configurations, $U'$ is not isomorphic (as a design) to the classical unital.

Furthermore, Grüning shows that $U'$ is isomorphic to its dual. Thus $U'$ is embeddable in two nonisomorphic planes, namely, the Hall plane and the dual Hall plane. We also note that in this case the set aff$U$ = aff$U'$ can be completed in two nonisomorphic ways to a unital; one is embeddable in the classical plane and one is embeddable in the Hall plane.

Now we consider the case when the classical unital $U$ is secant to $\ell_\infty$ and we derive with respect to a derivation set disjoint from $U \cap \ell_\infty$. This time the regulus corresponding to $U \cap \ell_\infty$ is itself a regulus of $\mathcal{S}'$, and hence the same nonsingular quadric $U$ "meets" $\mathcal{S}'$ in this regulus. Thus by Theorem 4.2 we know $U'$ is a nonsingular-Buekenhout unital of Hall($q^2$). This case was studied in detail by Barwick [34] (see also Rinaldi [187]).

**Theorem 5.2.** *Let $U$ be a classical unital in* PG($2, q^2$) *secant to $\ell_\infty$. If we derive* PG($2, q^2$) *with respect to a derivation set of $\ell_\infty$ disjoint from $U \cap \ell_\infty$, then $U'$ is a nonsingular-Buekenhout unital in* Hall($q^2$). *Furthermore, $U'$ does not contain O'Nan configurations with 2 or 3 points in $U' \cap \ell'_\infty$, but does contain O'Nan configurations if $q > 5$.*

See [34] for the proof of the existence of O'Nan configurations. In particular, it is shown that if $Q$ is a point in $U' \setminus \ell'_\infty$ and $\ell$ is a secant line of $U'$ through $Q$ that meets the derivation set, then there is an O'Nan configuration of $U'$ containing $Q$ and $\ell$. Hence this unital is not isomorphic to the classical unital. Further, by analyzing the O'Nan configurations, it is shown that if $q > 3$, this unital is not isomorphic to the unital constructed in Theorem 5.1.

Next we consider the case where the classical unital $U$ is secant to $\ell_\infty$, and we derive with respect to a derivation set $D$ that at least one and at most $q$ points in common with $U$. Note that as $U \cap \ell_\infty$ and $D$ are Baer sublines, they necessarily meet in 1 or 2 points in this case. This case was studied in [34].

**Theorem 5.3.** *Let $U$ be a classical unital in $PG(2, q^2)$ which is secant to $\ell_\infty$, and derive with respect to a derivation set $D$ of $\ell_\infty$ satisfying $0 < |U \cap D| < q + 1$. Then $U'$ is not a unital of $\text{Hall}(q^2)$. Further, if $q > 3$, there are no nonsingular-Buekenhout unitals in $\text{Hall}(q^2)$ that contain at least one and at most $q$ points of the derivation set $D'$.*

*Proof.* Let $\mathcal{R}$ be the regulus of $\mathcal{S}$ corresponding to $D$, and let $\mathcal{R}'$ be its opposite regulus. Thus $\mathcal{S}' = (\mathcal{S} \setminus \mathcal{R}) \cup \mathcal{R}'$ is the spread which constructs the Hall plane $\text{Hall}(q^2)$. From our assumption we know that precisely one or two of the lines in $\mathcal{R}$ lie on the nonsingular quadric $\mathcal{U}$ corresponding to $U$. Hence each line of the opposite regulus $\mathcal{R}'$ contains either one or two points of $\mathcal{U}$. Since the points of $D'$ can be identified with the lines of $\mathcal{R}'$, we see that $U'$ meets $\ell_\infty' = (\ell_\infty \setminus D) \cup D'$ in either $(q + 1) + q = 2q + 1$ or $(q + 1) + (q - 1) = 2q$ points. In either case, $U'$ is certainly not a unital in $\text{Hall}(q^2)$.

In fact, if $q > 3$, then every regulus contains at least 5 lines. Since $\mathcal{S}$ is a regular spread and every 3 mutually skew lines uniquely determine a regulus, we see that no regulus of $\mathcal{S}'$ can contain exactly one or two lines of $\mathcal{R}'$. Hence, if $q > 3$, then there are no nonsingular-Buekenhout unitals in $\text{Hall}(q^2)$ with at least one and at most $q$ points of the derivation set $D'$.     □

We next consider the case when the classical unital $U$ is tangent to $\ell_\infty$ in $PG(2, q^2)$. In fact, the results here are true for any ovoidal-Buekenhout-Metz unital tangent to $\ell_\infty$ in $PG(2, q^2)$. If we derive with respect to a derivation set that is disjoint from $U \cap \ell_\infty$, then $U'$ corresponds in the Bruck-Bose representation to an ovoidal cone that meets $\Sigma_\infty$ in a line of the spread $\mathcal{S}'$. Hence $U'$ is an ovoidal-Buekenhout-Metz unital in $\text{Hall}(q^2)$ by Theorem 4.2. This case was studied in detail by Barwick [35] (see also Rinaldi [188]).

**Theorem 5.4.** *Let $U$ be an ovoidal-Buekenhout-Metz unital with respect to $\ell_\infty$ in $PG(2, q^2)$. If we derive $PG(2, q^2)$ with respect to a derivation set of $\ell_\infty$ disjoint from $U \cap \ell_\infty$, then $U'$ is an ovoidal-Buekenhout-Metz unital of $\text{Hall}(q^2)$. Further, if $U$ is classical, then $U'$ contains no O'Nan configurations with $U' \cap \ell_\infty'$ as a vertex, but does contain O'Nan configurations if $q > 5$.*

See [35] for the proof of the existence of O'Nan configurations. In particular, it is shown that if $U$ is classical, if $Q$ is a point in $U' \setminus \ell_\infty'$, and if $\ell$ is a secant line of $U'$ through $Q$ that meets the derivation set, then there is an O'Nan configuration of $U'$ containing $Q$ and $\ell$. Hence, if $U$ is classical and $q > 5$, then $U'$ is not isomorphic to the classical unital; furthermore, it is not isomorphic to the unital constructed in Theorem 5.1 (which has an O'Nan configuration through every point of $U' \cap \ell_\infty'$). It is also shown in [35] that if $U$ contains O'Nan configurations, then so does $U'$. As mentioned at the end of Chapter 4, there are no known constructions of O'Nan configurations in nonclassical ovoidal-Buekenhout-Metz unitals which are embedded in $PG(2, q^2)$. Such constructions would be useful in studying the structure of $U'$ in the above case.

The final situation to consider is deriving an ovoidal-Buekenhout-Metz unital $U$ tangent to $\ell_\infty$ with respect to a derivation set of $\ell_\infty$ that contains $U \cap \ell_\infty$. This case was studied in [35].

**Theorem 5.5.** *If $U$ is an ovoidal-Buekenhout-Metz unital in* $PG(2, q^2)$ *tangent to $\ell_\infty$, and if we derive with respect to a derivation set $D$ of $\ell_\infty$ containing the point $U \cap \ell_\infty$, then $U'$ is not a unital of* $Hall(q^2)$.

*Proof.* Let $S' = (S \backslash R) \cup R'$ be the spread used to construct $Hall(q^2)$, as in the proof of Theorem 5.3. Then one line of the regulus $R$ lies on the ovoidal cone $U$, and hence each line of the opposite regulus $R'$ contains one point of $U$. This means that $U'$ meets $\ell'_\infty$ in $q + 1$ points, and therefore $|U'| = q^3 + q + 1$. In particular, $U'$ is not a unital in $Hall(q^2)$. □

These results allow us to investigate whether there are any Buekenhout unitals in $Hall(q^2)$ that are not inherited from unitals in $PG(2, q^2)$. Let $U$ be an ovoidal cone in $PG(4, q)$ that meets $\Sigma_\infty$ in a line $\ell$ of $R'$, that is, $\ell$ is a generator of $U$. Then $U$ corresponds to an ovoidal-Buekenhout-Metz unital of $Hall(q^2)$. By a similar argument to the proof of Theorem 5.5, the corresponding point set in $PG(2, q^2)$ is not a unital of $PG(2, q^2)$, that is, this unital is not inherited from a unital of $PG(2, q^2)$. Note that by using similar arguments to the above theorems, we can also deduce that this is the only Buekenhout unital in $Hall(q^2)$ that is not inherited from a Buekenhout unital in $PG(2, q^2)$.

We now have completely answered, among other things, the question of what happens to the classical unital when we derive the Desarguesian plane to obtain the Hall plane. There are still more cases to consider involving derivation when $U$ is a nonclassical ovoidal-Buekenhout-Metz unital. In particular, we assume that $U$ is an ovoidal-Buekenhout-Metz unital with respect to $P_\infty = U \cap \ell_\infty$ in $PG(2, q^2)$, and we derive with respect to a derivation set on some line $\ell$ other than $\ell_\infty$. In such a situation, the special case in which it is known that we do indeed obtain a unital in $Hall(q^2)$ is when we derive with respect to $D = U \cap \ell$ for some secant line $\ell$ through $P_\infty$ (see Figure 5.2).

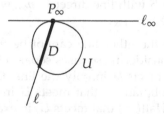

**Fig. 5.2.** Deriving the ovoidal-Buekenhout-Metz unital in $PG(2, q^2)$

This case was looked at by Dover [103], and it is also a special case of a construction given by Barlotti and Lunardon [31]. We follow the proof of

Dover. Note that since $U$ is an ovoidal-Buekenhout-Metz unital with respect to $P_\infty$, we know that $U \cap \ell$ is a Baer subline and hence is a derivation set.

**Theorem 5.6.** *Suppose that $U$ is an ovoidal-Buekenhout-Metz unital with respect to $P_\infty = U \cap \ell_\infty$ in $\mathrm{PG}(2, q^2)$. Let $\ell$ be a secant of $U$ through $P_\infty$ and derive $\mathrm{PG}(2, q^2)$ with respect to $D = U \cap \ell$. Then $U'$ is a unital of $\mathrm{Hall}(q^2)$.*

*Proof.* We will show that $U'$ is a set of $q^3 + 1$ points such that any line of $\mathrm{Hall}(q^2)$ meets $U'$ in 1 or $q + 1$ points. Note that we are deriving with respect to a derivation set $D$ that is not on $\ell_\infty$. The points of $\mathrm{Hall}(q^2)$ are the points of $\mathrm{PG}(2, q^2) \setminus D$, together with $q + 1$ points $D'$ on the line $\ell'$ of $\mathrm{Hall}(q^2)$. That is, the translation line of $\mathrm{Hall}(q^2)$ is $\ell' = (\ell \setminus D) \cup D'$. Hence the points of $\mathrm{aff}\, U = U \setminus \ell$ are the same as the points of $\mathrm{aff}\, U' = U' \setminus \ell'$ in $\mathrm{Hall}(q^2)$. Further, as $D = U \cap \ell$, the $q$-secants of $\mathrm{aff}\, U'$ all meet $\ell'$ in points of $D'$. Hence $U' = \mathrm{aff}\, U' \cup D'$ and thus $\ell'$ meets $U'$ in $q + 1$ points. Thus $U'$ has $q^3 + 1$ points.

The lines of the Hall plane distinct from $\ell'$ fall into two types: type I are lines of $\mathrm{PG}(2, q^2)$ that meet $\ell$ in a point not contained in $D$; type II are Baer subplanes of $\mathrm{PG}(2, q^2)$ that contain $D$. Lines of $\mathrm{PG}(2, q^2)$ that meet $\ell$ in a point not contained in $D$ meet $U$ in 1 or $q + 1$ points, hence a line of type I in $\mathrm{Hall}(q^2)$ meets $U'$ in 1 or $q + 1$ points.

We now consider lines of type II. We work in the Bruck-Bose representation of $\mathrm{PG}(2, q^2)$ with respect to $\ell_\infty$, and thus the points of $\ell_\infty$ correspond to lines of the regular spread $\mathcal{S}$ in $\Sigma_\infty$. Note, however, that the given derivation does not correspond to reversing a regulus in this spread. The unital $U$ corresponds to an ovoidal cone $\mathcal{U}$ that meets $\Sigma_\infty$ in the spread line $p_\infty$. The line $\ell$ corresponds to a plane $\pi_\ell$ of $\mathrm{PG}(4, q) \setminus \Sigma_\infty$ through $p_\infty$. The derivation set $D$ corresponds to a line $d$ of $\pi_\ell$ that meets $p_\infty$ in the vertex $V$ of $\mathcal{U}$.

A Baer subplane of $\mathrm{PG}(2, q^2)$ containing $D$ takes one of the following three forms in $\mathrm{PG}(4, q)$:

(a) a plane of $\mathrm{PG}(4, q)$ that meets $\mathcal{U}$ in only the line $d$,
(b) a plane of $\mathrm{PG}(4, q)$ that meets $\mathcal{U}$ in a pair of intersecting lines, one of which is $d$,
(c) a ruled cubic surface $\mathcal{B}$ with line directrix $p_\infty$, where $d$ is a generator of this surface and $\mathcal{B} \cap \Sigma_\infty = p_\infty$.

Note that in case (b) the other line cannot be $p_\infty$ since then the plane would be $\pi_\ell$. We now consider these cases separately. For case (a), let $\pi$ be a plane of $\mathrm{PG}(4, q)$ that meets $\mathcal{U}$ in only the line $d$. Hence in $\mathrm{PG}(2, q^2)$, $\pi$ corresponds to a Baer subplane $\mathcal{B}$ that meets $U$ in $D$. This Baer subplane corresponds to a line of $\mathrm{Hall}(q^2)$ that meets $U'$ in exactly one point, namely, the point of $D'$ corresponding to the slope point of $\mathcal{B}$. For case (b), let $\pi$ be a plane of $\mathrm{PG}(4, q)$ that meets $\mathcal{U}$ in two intersecting lines, one of which is $d$. So $\pi$ meets $\mathcal{U} \setminus D$ in $q$ collinear points. As $p_\infty$ is not one of the lines of $\mathcal{U}$ in $\pi$, these $q$ points are points of $\mathrm{PG}(4, q) \setminus \Sigma_\infty$. Hence in $\mathrm{PG}(2, q^2)$, the Baer subplane $\mathcal{B}$ corresponding to $\pi$ meets $U \setminus D$ in $q$ points. So when considered as a line of

Hall($q^2$), $\mathcal{B}$ contains $q$ points of $U'\backslash D'$ and one point of $D'$ (namely, the slope point of $\mathcal{B}$); that is, $|\mathcal{B} \cap U'| = q + 1$.

Finally, we consider case (c). Let $\mathcal{B}$ be a ruled cubic surface that meets $\Sigma_\infty$ in its line directrix $p_\infty$ and has the line $d$ as a line generator. Let $m_1, m_2, \ldots, m_q$ be the remaining $q$ line generators of $\mathcal{B}$. Note that $d$ is a line of $\mathcal{U}$ and thus contains the vertex $V$ of $\mathcal{U}$. Hence none of the other generators $m_1, m_2, \ldots, m_q$ contains $V$. Consider the planes $\pi_i = \langle p_\infty, m_i \rangle$ for $i = 1, 2, \ldots, q$. These planes are distinct as the line generators of $\mathcal{B}$ are disjoint. There are $q^2$ planes of $\mathrm{PG}(4,q)\backslash\Sigma_\infty$ that contain $p_\infty$, and there are $q^2$ lines (generators) of $\mathcal{U}$ distinct from $p_\infty$. Hence every plane of $\mathrm{PG}(4,q)\backslash\Sigma_\infty$ through $p_\infty$ meets $\mathcal{U}$ in $p_\infty$ and one further line. Thus each plane $\pi_i$ meets $\mathcal{U}$ in $p_\infty$ and one further line, which we denote by $\ell_i$. As $m_i$ does not contain $V$, and $\ell_i$ does contain $V$, $m_i \cap \ell_i$ is a point of $\pi_i$ not in $\Sigma_\infty$. Hence $m_i$ meets $\mathcal{U}$ in two points, namely, $m_i \cap \ell_i$ and $m_i \cap p_\infty$, but only one of these is in $\mathrm{PG}(4,q)\backslash\Sigma_\infty$. In $\mathrm{PG}(2,q^2)$, $\mathcal{B}$ corresponds to a Baer subplane $\mathcal{B}$ that contains $q$ points of $U\backslash D$ (one on each of the lines $m_i$). Hence as a line of Hall($q^2$), $\mathcal{B}$ meets $U'$ in $q$ points of $U'\backslash D'$ and one point in $D'$ (namely, the slope point of $\mathcal{B}$); that is, $|\mathcal{B} \cap U'| = q + 1$.

Thus we have shown that the lines of Hall($q^2$) corresponding to Baer subplanes of $\mathrm{PG}(2,q^2)$ (that is, lines of type II) all meet $U'$ in 1 or $q + 1$ points. Hence $U'$ is a unital of Hall($q^2$). □

We note that the existence of O'Nan configurations in these unitals is unknown in general. Again, constructions of O'Nan configurations in non-classical ovoidal-Buekenhout-Metz unitals in $\mathrm{PG}(2,q^2)$ would assist in analyzing O'Nan configurations in these unitals. If $U$ is classical, then $U$ is an orthogonal-Buekenhout-Metz unital with respect to any of its tangent lines, and $U$ is a nonsingular-Buekenhout unital with respect to any of its secant lines. Hence the unital $U'$ in Theorem 5.6 is isomorphic to the nonsingular-Buekenhout unital obtained in Theorem 5.1 when $U$ is classical. The next theorem shows that this is not the case if $U$ is nonclassical.

**Theorem 5.7.** *Suppose that $U$ is an ovoidal-Buekenhout-Metz unital with respect to $P_\infty = U \cap \ell_\infty$ in $\mathrm{PG}(2,q^2)$. Let $U'$ be the unital in Hall($q^2$) obtained by deriving $\mathrm{PG}(2,q^2)$ with respect to $D = U \cap \ell$ for some secant $\ell$ through $P_\infty$. Then $U'$ is a nonsingular-Buekenhout unital if and only if $U$ is classical. Thus, Hall($q^2$) contains unitals which are not Buekenhout unitals.*

*Proof.* This time we use the Bruck-Bose representation of $\mathrm{PG}(2,q^2)$ with respect to the line $\ell$, and thus the points of $\ell$ correspond to the lines of the regular spread $\mathcal{S}$ in $\Sigma_\infty$. Let $\mathcal{R}$ be the regulus of $\mathcal{S}$ corresponding to the derivation set $D$, and let $\mathcal{R}'$ be the opposite regulus. Then the spread $\mathcal{S}' = (\mathcal{S}\backslash\mathcal{R}) \cup \mathcal{R}'$ constructs the Hall plane Hall($q^2$). Note that $\ell' = (\ell \backslash D) \cup D'$ is the translation line of Hall($q^2$). We showed in Theorem 5.1 that if $U$ is classical, then $U'$ is a nonsingular-Buekenhout unital. Conversely, suppose that $U'$ is a nonsingular-Buekenhout unital in Hall($q^2$), and hence $U'$ corresponds to a nonsingular quadric $\mathcal{U}'$ that "meets" $\mathcal{S}'$ in the regulus $\mathcal{R}'$. Then in the

Bruck-Bose model for $PG(2, q^2)$, $U'$ is a nonsingular quadric that "meets" the spread $S$ in the regulus $\mathcal{R}$, and hence $U$ is a nonsingular-Buekenhout unital in $PG(2, q^2)$. By Theorem 4.5, $U$ is classical.                               □

Dover [103] also investigated the equivalences among these unitals in Hall$(q^2)$. In particular, suppose that $U_1$ and $U_2$ are *orthogonal*-Buekenhout-Metz unitals of $PG(2, q^2)$ with respect to $P_\infty$ such that $U_1 \cap \ell = U_2 \cap \ell$ for some line $\ell$, other than $\ell_\infty$, through $P_\infty$. If one derives with respect to the derivation set $D = U_1 \cap \ell = U_2 \cap \ell$, then it is shown in [103] that $U_1'$ and $U_2'$ are equivalent in Hall$(q^2)$ if and only if $U_1$ and $U_2$ are equivalent in $PG(2, q^2)$.

It should be noted that in all constructions given in this section, the point set is kept fixed (the given unital $U$ embedded in $PG(2, q^2)$) and the model for the Hall plane is changed by deriving $PG(2, q^2)$ in different ways. Sometimes the given point set turns out to be a unital in the Hall plane, and sometimes it does not. This is quite different than the approach outlined in Chapter 4, which was used to classify the nonsingular and orthogonal-Buekenhout-Metz unitals embedded in the Desarguesian plane $PG(2, q^2)$. Namely, in the approach in Chapter 4, the model for the translation plane is fixed (by fixing the spread $S$ of $\Sigma_\infty$ in the Bruck-Bose representation), and then different point sets (ovoidal cones and parabolic quadrics) are used to create different unitals embedded in the given plane. One could also use this approach to construct, and perhaps classify, the Buekenhout unitals embedded in the Hall plane.

Fixing a Hall spread $S' = (S \setminus \mathcal{R}) \cup \mathcal{R}'$ in $\Sigma_\infty \cong PG(3, q)$, exhaustive searching using the software package Magma [84] yielded five inequivalent Buekenhout unitals in the Hall plane of order 16 and seven inequivalent Buekenhout unitals in the Hall plane of order 25. In Hall(16) there are three inequivalent orthogonal-Buekenhout-Metz unitals (two of which arise from Theorem 5.4) and two inequivalent nonsingular-Buekenhout unitals (one each from Theorems 5.1 and 5.2). The remaining orthogonal-Buekenhout-Metz unital meets the line at infinity in a point corresponding to a line of $\mathcal{R}'$ (this case is discussed after Theorem 5.5). In Hall(25) there are four inequivalent orthogonal-Buekenhout-Metz unitals (three arising from Theorem 5.4) and three inequivalent nonsingular-Buekenhout unitals (one arising from Theorem 5.1 and two from Theorem 5.2). Again the remaining orthogonal-Buekenhout-Metz unital in Hall(25) arises as in Hall(16).

It seems clear that there are more inequivalent unitals embedded in the Hall plane of order $q^2$ than in the Desarguesian plane of order $q^2$. Of course, part of the reason is our definition of equivalence and the fact that the Desarguesian plane has a much larger automorphism group than the Hall plane. It is an interesting open question whether inequivalent unitals embedded in the Hall plane (or any given plane) remain inequivalent (actually, nonisomorphic) when treated solely as designs.

It should be noted that the Grüning unitals have full stabilizers in the automorphism group of the Hall plane that are isomorphic to the full stabilizer of the Hall spread (and thus of order 1200 for $q = 4$ and order 1440 for $q = 5$). In particular, the stabilizer of the Grüning unital has one orbit of size $q + 1$ and one orbit of size $q^3 - q$ when acting on the points of the unital. In [127] it is shown that the Grüning unital and classical unital have isomorphic groups when treated as designs. The other Buekenhout unitals found in Hall(16) and Hall(25) have relatively small groups with no particularly interesting group action.

We close this section by mentioning an alternative approach to determining when unitals in derivable planes give rise to unitals in the derived planes. This approach was developed by Blokhuis and O'Keefe [57], and their results can be used to reproduce the above constructions of unitals in the Hall plane. We state these results below. The proofs, which we omit, rely on the blocking set characterization of Bruen and Thas [80] given in Theorem 2.16.

**Theorem 5.8.** *Let $\mathcal{P}$ be a projective plane of order $q^2$. Let $D$ be a derivation set of $\ell_\infty$, and let $U$ be a unital of $\mathcal{P}$. Let $U'$ be the set of points in the derived plane $\mathcal{D}(\mathcal{P})$ corresponding to $U$.*

1. *If $U \cap D = D$ and each Baer subplane containing $D$ meets $U$ in $q + 1$ or $2q + 1$ points, then $U'$ is a unital of $\mathcal{D}(\mathcal{P})$.*
2. *Suppose that $U \cap D = D$, each Baer subplane of $\mathcal{P}$ containing $D$ meets $U$ in at least $q + 1$ points, and there are at least $q + 1$ Baer subplanes on $D$ meeting $U$ in exactly $q + 1$ points. Then $U'$ is a unital in $\mathcal{D}(\mathcal{P})$.*
3. *If $U \cap D = \varnothing$ and each Baer subplane of $\mathcal{P}$ containing $D$ meets $U$ in 1 or at least $q + 1$ points, then $U'$ is a unital of $\mathcal{D}(\mathcal{P})$.*
4. *Suppose that $U \cap D = \varnothing$ and each Baer subplane of $\mathcal{P}$ containing $D$ meets $U$ in at least one point. Furthermore, assume that*
   (i) *if $\ell_\infty$ is a secant of $U$, then at least $(q + 1)^2$ Baer subplanes on $D$ meet $U$ in exactly one point, and*
   (ii) *if $\ell_\infty$ is a tangent of $U$, then at least $q(q + 1)$ Baer subplanes on $D$ meet $U$ in exactly one point.*
   *Then $U'$ is a unital of $\mathcal{D}(\mathcal{P})$.*

## 5.2 Unitals in Semifield Planes

Any translation plane admitting a polarity must be a dual translation plane as well. In particular, the algebraic system used to coordinatize the plane (see [140]) must satisfy both the left and the right distributive law, as well as several other properties associated with being a translation plane. More precisely, this forces the coordinatizing algebraic system to be a **semifield**, in which case the resulting (translation) plane is called a **semifield plane**. Basically, a semifield differs from a field in that the multiplication need not be associative or commutative. A necessary and sufficient condition for a translation plane $\mathcal{P}(\mathcal{S})$ to be a semifield plane is that the associated spread $\mathcal{S}$

admits a group which fixes one spread element pointwise and acts regularly on the remaining spread elements (see Theorem 1.3 in [27], for instance).

While not every semifield plane admits a polarity, it is possible to write down necessary and sufficient conditions (in terms of the existence of certain mappings on the coordinatizing semifield) for polarities to exist (see [120]). In particular, the following result may be found in Ganley [120].

**Theorem 5.9.** *If $\mathcal{D}$ is a finite commutative semifield which has a nontrivial involutory automorphism, then the associated translation plane $\mathcal{P}(\mathcal{D})$ admits a unitary polarity.*

In the above theorem, $\mathcal{P}(\mathcal{D})$ denotes the projective plane coordinatized by the semifield $\mathcal{D}$. This will be precisely defined below.

Now let $F = \mathrm{GF}(q)$ denote the (unique) finite field of order $q$, where $q$ is an odd prime power which is not a prime. Let $w$ be a nonsquare in $F$, and let $\alpha$ be some nonidentity automorphism of $F$ (for instance, the Frobenius automorphism). We define a new algebraic structure $\mathcal{D}$, whose elements are the ordered pairs $\{\langle x,y \rangle \mid x,y \in F\}$. Addition in $\mathcal{D}$ is defined componentwise, using the addition of $F$ in each component. However, multiplication in $\mathcal{D}$ is defined by the following rule:

$$\langle x_1, y_1 \rangle \circ \langle x_2, y_2 \rangle = \langle x_1 x_2 + w y_1^\alpha y_2^\alpha, \; y_1 x_2 + x_1 y_2 \rangle.$$

It turns out that $\mathcal{D}$ together with the above two binary operations of addition and multiplication is a commutative semifield of order $q^2$, called the **Dickson semifield**, which is not a field. Moreover, the mapping

$$\phi : \langle x,y \rangle \mapsto \langle x, -y \rangle$$

is an involutory automorphism of $\mathcal{D}$. Hence from Theorem 5.9 we know that $\mathcal{P}(\mathcal{D})$, the associated semifield plane of order $q^2$, admits a unitary polarity.

The points of $\mathcal{P}(\mathcal{D})$ are

$$\{(c,d) \mid c,d \in \mathcal{D}\} \cup \{(m) \mid m \in \mathcal{D}\} \cup \{(\infty)\},$$

where $\infty$ is some formal symbol which is not in $\mathcal{D}$. Recall that, as described above, the elements of $\mathcal{D}$ are actually ordered pairs of elements of $F$. The translation line (line at infinity) for $\mathcal{P}(\mathcal{D})$ consists of the points

$$\{(m) \mid m \in \mathcal{D}\} \cup \{(\infty)\},$$

where we think of $(m)$ as being the point lying on all lines whose *slope* is $m$ and we think of $(\infty)$ as being the point lying on all *vertical* lines. In fact, the lines of $\mathcal{P}(\mathcal{D})$ are

$$\{[m,k] \mid m,k \in \mathcal{D}\} \cup \{[k] \mid k \in \mathcal{D}\} \cup \{[\infty]\},$$

where

$$[m,k] = \{(m)\} \cup \{(c,d) \mid c,d \in \mathcal{D} \text{ with } m \circ c + d = k\},$$
$$[k] = \{(\infty)\} \cup \{(k,d) \mid d \in \mathcal{D}\},$$
$$[\infty] = \{(\infty)\} \cup \{(m) \mid m \in \mathcal{D}\}.$$

That is, the line $[m,k]$ has $-m$ as its "slope" and $k$ as its "y-intercept", the line $[k]$ is "vertical" with "x-intercept" $k$, and $[\infty]$ is the line at infinity (the translation line).

The unitary polarity of $\mathcal{P}(\mathcal{D})$ guaranteed by Theorem 5.9 is the mapping

$$\rho : (c,d) \leftrightarrow [c^\phi, -d^\phi],$$
$$(m) \leftrightarrow [m^\phi],$$
$$(\infty) \leftrightarrow [\infty].$$

The absolute points of this unitary polarity, as shown in [120], are

$$U = U_{q,\alpha} = \{(\infty)\} \cup \{(\langle x_1, y_1 \rangle, \langle -(x_1^2 - wy_1^{2\alpha})/2, y_2 \rangle) \mid x_1, y_1, y_2 \in F\}.$$

For $q = 9$ the software package Magma [84] was used to compute the Dickson semifield plane $\mathcal{P}(\mathcal{D}_{81})$ of order 81 and the unitary polarity described above, using the Frobenius map $\sigma : x \mapsto x^3$ as the nontrivial field automorphism $\alpha$ of GF(9). We let $U_9$ denote the resulting (Ganley) unital in $\mathcal{P}(\mathcal{D}_{81})$. The full automorphism group of this translation plane has order $2^9 3^{12}$, and the full stabilizer $G$ of $U_9$ has order $2^6 3^6$. All Sylow subgroups of $G$ are non-abelian, and $G$ is the semidirect product of its unique Sylow 3-subgroup $S_3$ by one of its Sylow 2-subgroups. In fact, the subgroup $S_3$ (of order $3^6$) fixes the point $(\infty)$, acts transitively on the points of $[\infty] \setminus (\infty)$, and acts semiregularly on the points of $\mathcal{P}(\mathcal{D}_{81}) \setminus [\infty]$. In particular, $S_3$ acts sharply transitively on the points of $U_9 \setminus (\infty)$. Of course, $[\infty]$ is the unique tangent line to $U_9$ at $(\infty)$. As the points of $U_9$ are the absolute points of a unitary polarity, the feet of any point of $\mathcal{P}(\mathcal{D}_{81}) \setminus U_9$ are collinear.

The translations of $\mathcal{P}(\mathcal{D}_{81})$ which leave $U_9$ invariant are precisely the 9 elations, including the identity, contained in $S_3$ and forming an elementary abelian subgroup of $S_3$. There are 81 nonidentity affine homologies of $\mathcal{P}(\mathcal{D}_{81})$ which leave $U_9$ invariant; the axes for these homologies are the 81 lines other than $[\infty]$ passing through $(\infty)$. The centers of these homologies are all distinct and comprise the 81 points of $[\infty]$ other than $(\infty)$. All these homologies have order 2.

It is currently unknown if the group actions described above for $U_9$ generalize to arbitrary $U_{q,\alpha}$.

Since the Dickson semifield plane $\mathcal{P}(\mathcal{D})$ is also a two-dimensional translation plane, it contains ovoidal-Buekenhout-Metz unitals as discussed in Section 4.1. Again for $q = 9$, the software package Magma [84] may be used to construct the Buekenhout unitals embedded in $\mathcal{P}(\mathcal{D}_{81})$. Exhaustive searching found 26 pairwise inequivalent orthogonal-Buekenhout-Metz unitals and one nonsingular-Buekenhout unital in $\mathcal{P}(\mathcal{D}_{81})$.

Thirteen of the orthogonal-Buekenhout-Metz unitals arise from an elliptic cone whose unique spread line generator is the line of the Dickson semifield spread that is fixed pointwise, as described above. These unitals have full stabilizers of order $2^1 3^6$, $2^2 3^6$, or $2^3 3^6$, and all such stabilizers fix the point $(\infty)$ and act transitively on the remaining points of the unital. In fact, in all these cases the Sylow 3-subgroup is a normal, non-abelian subgroup which acts sharply transitively on the affine points of the unital. The elations and homologies leaving these unitals invariant are the same ones as described above for the Ganley unital. The only points not lying on one of these unitals which have collinear feet are the points of $[\infty] \setminus (\infty)$.

The remaining 13 ovoidal-Buekenhout-Metz unitals in $\mathcal{P}(\mathcal{D}_{81})$ arise from elliptic cones whose unique spread line generator is not the spread line that is fixed pointwise. Thus these unitals are tangent to $[\infty]$ at some point other than $(\infty)$. These unitals have much smaller stabilizers, of order 18, 36, or 72. In all cases the Sylow 3-subgroup of order 9 consists of all the translations leaving the unital invariant. In each case there is a unique affine homology, whose order is 2, leaving the unital invariant. For two of the unitals there are 99 points off the unital with collinear feet. In the other cases, the points off the unital with collinear feet are precisely the points of $[\infty]$ other than the point of tangency.

The unique nonsingular-Buekenhout unital of $\mathcal{P}(\mathcal{D}_{81})$, up to equivalence, has a stabilizer of order $2^6 3^2$. There are 297 points off this unital which have collinear feet, and there are no affine homologies leaving the unital invariant. The elations leaving the unital invariant are the usual ones described above.

The main reason for our concern about the homologies leaving the above unitals invariant is the following theorem of Abatangelo, Korchmáros, and Larato [4].

**Theorem 5.10.** *Let $\mathcal{P}$ be a translation plane of odd order containing a transitive parabolic unital $U$ (that is, a unital meeting the line at infinity in one point and admitting a stabilizer which acts transitively on the affine points of the unital). Assume the collineation group $G$ of $\mathcal{P}$ leaving $U$ invariant contains an affine homology. Then $\mathcal{P}$ is a semifield plane, and $G$ has a normal subgroup $K$ that acts on the affine points of $U$ as a sharply transitive permutation group.*

The Ganley unital and the first thirteen orthogonal-Buekenhout-Metz unitals described above in the commutative semifield plane $\mathcal{P}(\mathcal{D}_{81})$ are of this type, while the theorem does not apply to the other fourteen Buekenhout unitals in $\mathcal{P}(\mathcal{D}_{81})$.

In [4] unitals embedded in square order semifield planes coordinatized by **Albert twisted fields** are described (see [11] for the original description of these semifields). In particular, it is shown that there is a transitive parabolic unital stabilized by an affine homology in any such square order Albert twisted field plane. These unitals can be constructed as a union of ovals which pairwise meet at a point on the line at infinity, in complete analogy to the special orthogonal-Buekenhout-Metz unitals embedded in $PG(2, q^2)$, for odd $q$, which are described in Theorem 4.19 of Section 4.2.1. The description given in [4] is purely algebraic, and it is currently unknown if these unitals are, in fact, ovoidal-Buekenhout-Metz unitals. The smallest possible square order for such an Albert twisted field plane is $3^6$, and this is too large for an in-depth Magma computation, at least with our present technology.

The Albert twisted fields can be described as follows. Let $q$ be an odd prime power, and suppose $GF(q)$ contains a subfield $GF(d)$ such that $-1$ is not a $(d - 1)$ power in $GF(q)$. Then every element in $GF(q)$ can be uniquely expressed as $x = a^d + a$, for some $a \in GF(q)$. The Albert twisted field of order $q$, which is a commutative semifield, can then be obtained from $GF(q)$ by using the usual addition and defining multiplication as follows: $(a^d + a) \circ (b^d + b) = a^d b + ab^d$. If, in addition, $q$ is square, then one can show that this commutative semifield has an involutory automorphism, and hence by Theorem 5.9 the associated translation plane has a unitary polarity. In [3] the unitals obtained as the absolute points of such a unitary polarity are carefully studied, and the stabilizer subgroups are completely determined. By comparing the stabilizer subgroups of these unitals with the stabilizer subgroups of the unitals which are unions of ovals, it is shown that these two families of unitals in the Albert twisted field planes are indeed inequivalent.

## 5.3 Unitals in Nearfield Planes

In this brief section we discuss unitals in two-dimensional translation planes which are coordinatized by *nearfields*. A **nearfield** is an algebraic system that differs from a field in that the multiplication need not be commutative and only one distributive law is assumed. In relation to semifields, you gain associativity of multiplication but lose one of the distributive laws. If the right distributive law is assumed, then the nearfield is called a **right nearfield**. As with semifields, the translation planes coordinatized by nearfields are called **nearfield planes**.

Here we are concerned with a family of two-dimensional nearfield planes, called *regular*. The **regular nearfield planes** are a special case of André planes, as defined in Section 3.4.3. In particular, for such a translation plane the corresponding spread of $PG(3, q)$, where $q$ is necessarily odd, is obtained by reversing a carefully chosen linear set of $(q - 1)/2$ disjoint reguli in a regular spread. Alternatively, these planes can be coordinatized, as discussed in the last section, by starting with a certain algebraic structure called a *regular nearfield* (see [96]). Specifically, start with the elements of the finite field $GF(q^2)$, where $q$ is any odd prime power (possibly an odd prime). We define addition as the usual field addition, but define multiplication as follows:

$$a \circ b = \begin{cases} ab & \text{if } b \text{ is a square (possibly 0) in } GF(q^2) \\ a^q b & \text{if } b \text{ is a non-square in } GF(q^2). \end{cases}$$

Here juxtaposition of elements means the usual multiplication in $GF(q^2)$. One can easily check that the set of elements in $GF(q^2)$ under the above two binary operations form a right nearfield, called a **regular (right) nearfield**. These algebraic structures, in a more general form, were discovered by Dickson [99].

If $\mathcal{N}$ is such a right nearfield, then one obtains the associated translation plane $\mathcal{P}(\mathcal{N})$ by defining the points to be

$$\{(c,d) \mid c,d \in \mathcal{N}\} \cup \{(m) \mid m \in \mathcal{N}\} \cup \{(\infty)\},$$

where $\infty$ is some formal symbol which is not in $\mathcal{N}$. As discussed for semifield planes in Section 5.2, the translation line (line at infinity) for $\mathcal{P}(\mathcal{N})$ consists of the points

$$\{(m) \mid m \in \mathcal{N}\} \cup \{(\infty)\},$$

where we think of $(m)$ as being the point lying on all lines whose *slope* is $m$ and we think of $(\infty)$ as being the point lying on all *vertical* lines. In fact, the lines of $\mathcal{P}(\mathcal{D})$ are

$$\{[m,k] \mid m,k \in \mathcal{N}\} \cup \{[k] \mid k \in \mathcal{N}\} \cup \{[\infty]\},$$

where

$$[m,k] = \{(m)\} \cup \{(c,d) \mid c,d \in \mathcal{N} \text{ with } c \circ m + d = k\},$$
$$[k] = \{(\infty)\} \cup \{(k,d) \mid d \in \mathcal{N}\},$$
$$[\infty] = \{(\infty)\} \cup \{(m) \mid m \in \mathcal{N}\}.$$

Since we are coordinatizing by a right nearfield, rather than a left nearfield, we put the "slope" $m$ on the right in the above equation for the line $[m,k]$.

Using these coordinates, Wantz [222] described some orthogonal-Buekenhout-Metz unitals in $PG(2,q^2)$ that remain unitals in the regular nearfield planes of order $q^2$. Namely, letting $\epsilon = \zeta^{(q+1)/2}$ for some primitive element $\zeta$ of $GF(q^2)$, it is shown that

$$U_{(a,b)} = \{(x, ax^2 + bx^{q+1} + t\epsilon, 1) \mid x \in GF(q^2), t \in GF(q)\} \cup \{(0,1,0)\}$$

is such a unital for any $a,b \in GF(q)$ such that $b^2 - a^2$ is a nonzero square in $GF(q)$. The advantage of using these orthogonal-Buekenhout-Metz unitals, as opposed to the ones described in Section 4.2.1, is that these unitals are left invariant by the conjugation map $\sigma : x \mapsto x^q$. In [222] collineation groups stabilizing these unitals are computed and projective equivalences are sorted out. In particular, it is shown that for odd primes $q$ the number of such mutually inequivalent unitals constructed is $(q+1)/4$ if $q \equiv 3 \pmod 4$ and $(q+3)/4$ if $q \equiv 1 \pmod 4$. One of these equivalence classes generalizes the unital embedded in the Hall plane of order 9 found in the computer search of Brouwer [65]. In this regard it should be pointed out that the Hall plane of order 9 is also a regular nearfield plane, called the **exceptional nearfield plane**.

For $q = 5$ the software package Magma [84] was used to search exhaustively for Buekenhout unitals in the regular nearfield plane $\mathcal{P}(\mathcal{N}_{25})$ of order 25. In addition to the two Wantz unitals in this plane, the exhaustive search found seven other orthogonal-Buekenhout-Metz unitals and two nonsingular-Buekenhout unitals embedded in $\mathcal{P}(\mathcal{N}_{25})$. Perhaps the most interesting characteristic found for these unitals is the number of points which have collinear feet. For instance, one of the Wantz unitals (corresponding to the classical unital of $PG(2, 25)$) has 45 points off the unital with collinear feet. In addition to the points of $[\infty] \setminus (\infty)$, the 20 points off the unital on one of the secant lines through $(\infty)$ all have collinear feet. The full stabilizer of this unital in the automorphism group of $\mathcal{P}(\mathcal{N}_{25})$ has order 240, and it partitions the points of the unital into orbits of size 1, 5 and 120. The orbits of size 1 and 5 form a chord of the unital, whose extended line contains the other 20 points with collinear feet mentioned above.

## 5.4 Unitals Embedded in Nontranslation Planes

In this section we discuss some unitals embedded in square order projective planes which are not translation planes. Some of these unitals consist of the absolute points of associated polarities defined on the ambient planes, and some do not. We begin by recalling the result of Seib [200] concerning polarities in arbitrary square order projective planes, namely, Theorem 1.34.

**Theorem 5.11.** *Let* $\mathcal{P}$ *be any projective plane of order* $n^2$, *and suppose* $\rho$ *is a polarity defined on* $\mathcal{P}$. *Then the number of absolute points of* $\rho$ *is at most* $n^3 + 1$. *Moreover, if this bound is achieved, then the set of absolute points forms a unital embedded in* $\mathcal{P}$.

### 5.4.1 Figueroa Plane

The first family of nontranslation planes we discuss are the finite **Figueroa planes**. The order of these planes is necessarily the cube of a prime power. Since we are interested in embedded unitals in these planes, we require the order to be a square. Thus we will be interested in Figueroa planes of order $q^6$, where $q$ is a prime power. The construction we give for these planes is based on the synthetic description given by Grundhöfer in [125], not the original description given by Figueroa [117].

We begin with the canonical unitary polarity $\rho$ of the classical projective plane $PG(2, q^6)$ of order $q^6$. Thus, using homogeneous point and line coordinates for $PG(2, q^6)$,

$$(x, y, z)^\rho = \left( x^{q^3}, y^{q^3}, z^{q^3} \right)'.$$

Hence the absolute points of the unitary polarity $\rho$ of $PG(2,q^6)$ form the canonical Hermitian curve $\mathcal{H}$. Now let $\alpha$ be the automorphic collineation of $PG(2,q^6)$ of order 3 defined by

$$(x,y,z)^\alpha = \left( x^{q^2}, y^{q^2}, z^{q^2} \right).$$

Then the points and lines of $PG(2,q^6)$ fall into three disjoint classes each, according to the structure of the corresponding orbit under the cyclic group generated by $\alpha$. That is, we say that a point $P$ of $PG(2,q^6)$ is of type I if $P^\alpha = P$; or type II if $\{P, P^\alpha, P^{\alpha^2}\}$ consists of three distinct collinear points; or type III if $\{P, P^\alpha, P^{\alpha^2}\}$ consists of three noncollinear points. We dually define what it means for a line of $PG(2,q^6)$ to be of type I, type II, or type III. We now define an involutory bijection $\mu$ between the points of type III and the lines of type III as follows: if a point $P$ and a line $\ell$ are both of type III, then

$$P^\mu = P^\alpha P^{\alpha^2}, \qquad \ell^\mu = \ell^\alpha \cap \ell^{\alpha^2}.$$

Let $\mathcal{P}$ be the set of points of $PG(2,q^6)$, and let $\mathcal{L}$ be the set of lines of $PG(2,q^6)$. We define a new incidence, which we call $\mathcal{F}$-incidence, between $\mathcal{P}$ and $\mathcal{L}$. Let $P \in \mathcal{P}$ and let $\ell \in \mathcal{L}$. If $P$ and $\ell$ are both of type III, we say that $P$ is $\mathcal{F}$-incident with $\ell$ if and only if $\ell^\mu \in P^\mu$ in $PG(2,q^6)$. In all other cases, we say $P$ is $\mathcal{F}$-incident with $\ell$ if and only if $P \in \ell$ in $PG(2,q^6)$. As shown in [125], this incidence structure is the Figueroa projective plane of order $q^6$, which we denote by $\text{Fig}(q^6)$. It should be noted that this nonclassical plane is not a translation plane.

Unitals were constructed in $\text{Fig}(q^6)$ for all prime powers $q$ by de Resmini and Hamilton [95], and we follow their approach.

**Lemma 5.12.** *Let $q$ be any prime power, and let $\text{Fig}(q^6)$ be the Figueroa plane of order $q^6$ obtained from $PG(2,q^6)$ as above. Then the following are true.*

1. *Any point $P$ has the same type as the line $P^\rho$.*
2. *Any line $\ell$ has the same type as the point $\ell^\rho$.*
3. *For any point $P$ of type III, we have $P^{\rho\mu} = P^{\mu\rho}$.*
4. *For any line $\ell$ of type III, we have $\ell^{\rho\mu} = \ell^{\mu\rho}$.*

*Proof.* Parts (1) and (2) follow directly from the fact that the collineation $\alpha$ and the polarity $\rho$ commute. If $P$ is a point of type III, then

$$P^{\mu\rho} = \left( P^\alpha P^{\alpha^2} \right)^\rho = P^{\alpha\rho} \cap P^{\alpha^2\rho} = P^{\rho\alpha} \cap P^{\rho\alpha^2} = P^{\rho\mu},$$

proving (3). The proof of (4) follows similarly.                    □

**Theorem 5.13.** *For any prime power $q$, the Figueroa plane $\text{Fig}(q^6)$ of order $q^6$ admits a unitary polarity, and hence contains a unital.*

*Proof.* We show that the polarity $\rho$ of $PG(2, q^6)$ is also a polarity of $Fig(q^6)$. Since $\rho$ has order 2 as a mapping, it suffices to show that $\rho$ is a correlation on $Fig(q^6)$. Moreover, since the incidence in $Fig(q^6)$ and $PG(2, q^6)$ agree except when points and lines are both of type III, we may assume that $P$ is a point of type III and $\ell$ is a line of type III. Then $P^\rho$ and $\ell^\rho$ are both of type III by Lemma 5.12. Now $\ell^\rho$ is $\mathcal{F}$-incident with $P^\rho$ if and only if $P^{\rho\mu} = P^{\mu\rho}$ lies on $\ell^{\rho\mu} = \ell^{\mu\rho}$ in $PG(2, q^6)$, again using Lemma 5.12. Since $\rho$ is a polarity of $PG(2, q^6)$, this latter statement is equivalent to $\ell^\mu \in P^\mu$, which is the definition of $P$ being $\mathcal{F}$-incident with $\ell$. Hence $\rho$ is a correlation, and hence a polarity, of $Fig(q^6)$ by definition.

Next we count the number of absolute points of $\rho$ as a polarity of $Fig(q^6)$. A point $P \in Fig(q^6)$ is $\rho$-absolute if and only if $P$ is $\mathcal{F}$-incident with $P^\rho$, which means $P \in P^\rho$ if $P$ is of type I or II, and means $P^{\rho\mu} \in P^\mu$ if $P$ is of type III. In the former case, this simply means that $P$ is a type I or type II point of the Hermitian curve $\mathcal{H}$. In the latter case, applying Lemma 5.12 one more time, this means that $P^\mu$ is a type III tangent line to the Hermitian curve $\mathcal{H}$. But $P^\mu$ is a type III tangent line to $\mathcal{H}$ if and only if $P^{\mu\rho}$ is a type III point of $\mathcal{H}$. Hence, considering both cases and using the fact that $\mu\rho$ is a bijection, the number of $\rho$-absolute points of $Fig(q^6)$ is the same as the total number of points of $\mathcal{H}$, namely, $q^9 + 1$. Therefore, by Theorem 5.11, $\rho$ is a unitary polarity of $Fig(q^6)$, and the set of absolute points of this polarity forms a unital embedded in $Fig(q^6)$. □

## 5.4.2 Hughes Plane

The next projective plane we consider is the **Hughes plane** of order $q^2$, for any odd prime power $q$. Although this plane is self-dual, it does not admit unitary polarities (see [179]). Hence any unital that may be embedded in the Hughes plane will not arise as the set of absolute points for some polarity. We give the description for the Hughes plane presented in [190] by Rosati, not the original one given by Hughes [139]. Nonetheless, any description of the Hughes plane requires some algebra.

Let $F = GF(q^2)$ denote the finite field of odd order $q^2$, and let $K = GF(q)$ be its subfield of order $q$. Let $F_S$ and $F_N$ denote the nonzero squares and nonsquares in $F$, respectively. We let $\mathcal{N}$ be the regular (left) nearfield obtained from $F$ just as the regular (right) nearfield was obtained in Section 5.3. Namely, the elements of $\mathcal{N}$ are the elements of $F$. Addition in $\mathcal{N}$ is the usual field addition, but multiplication is defined as follows:

$$a \circ b = \begin{cases} ab & \text{if } a \in F_S \\ ab^q & \text{if } a \in F_N \\ 0 & \text{if } a = 0. \end{cases}$$

Notice that the conjugation automorphism $\sigma$ of $F$, where $\sigma : x \mapsto x^q$, is also an automorphism of $\mathcal{N}$. We use left regular nearfields instead of right regular nearfields in order to be consistent with the results found in [221], many of which will be discussed below.

Let $V$ be a three-dimensional vector space over $F$. We consider the vectors of $V^* = V \setminus \{(0,0,0)\}$ as three-tuples over $\mathcal{N}$, as well. We define an equivalence relation on $V^*$ by saying two nonzero vectors $P$ and $Q$ are equivalent if and only if $Q = a \circ P$ for some $a \in \mathcal{N}^* = \mathcal{N} \setminus \{0\}$, where the notation $a \circ P$ means $(a \circ x, a \circ y, a \circ z)$ if $P = (x,y,z) \in V^*$. Thus $V^*$ is partitioned into $q^4 + q^2 + 1$ equivalence classes, each of size $q^2 - 1$. Moreover, one may choose a unique left-normalized (or right-normalized) representative for each of these classes.

We now define an incidence relation on $V^*$ as follows. We choose the basis $\{1, \epsilon\}$ for $F = GF(q^2)$ over $K = GF(q)$, as before, where $\epsilon = \zeta^{(q+1)/2}$ for some primitive element $\zeta$ of $F$. If $P = (x,y,z)$ and $Q = (\alpha, \beta, \gamma)$ are nonzero vectors in $V^*$ with $\alpha = a_0 + a_1\epsilon$, $\beta = b_0 + b_1\epsilon$, and $\gamma = c_0 + c_1\epsilon$, for $a_0, a_1, b_0, b_1, c_0, c_1 \in K$, then we say $P$ and $Q$ are $\mathcal{H}$-incident if and only if

$$xa_0 + yb_0 + zc_0 + (xa_1 + yb_1 + zc_1) \circ \epsilon = 0. \qquad (5.1)$$

It turns out that this definition also induces an incidence relation on the equivalence classes of $V^*$ defined above.

Consider the incidence structure whose "points" and "lines" are both equal to the equivalence classes of $V^*$ defined above, and whose incidence is induced by equation (5.1). As shown in [190], this structure is a projective plane of order $q^2$, namely, the Hughes plane, which we will denote by $\mathrm{Hgh}(q^2)$. It should be emphasized that in this model the left-normalized (right-normalized) vectors in $V^*$ serve as representatives for the points (lines) in both the classical plane and the Hughes plane of order $q^2$.

A proof of the following useful lemma can be found in [221] (see [154] for an earlier version).

**Lemma 5.14.** *Let $S$ be a set of left-normalized vectors which represents a point set in both $PG(2, q^2)$ and $\mathrm{Hgh}(q^2)$, for any odd prime power $q$. Suppose that $S$ is invariant under the collineation induced by the conjugation mapping $\sigma$. If $\ell$ is any right-normalized vector representing a line of both $PG(2, q^2)$ and $\mathrm{Hgh}(q^2)$, then $\ell \cap S$ has the same cardinality independent of the plane in which the points and line are viewed.*

The following result of Wantz [221] then follows immediately.

**Theorem 5.15.** *Suppose the collineation induced by the conjugation automorphism leaves a set $S$ of left-normalized vectors invariant. Then $S$ represents a unital in the Hughes plane if and only if $S$ represents a unital in the classical plane.*

Rosati [191] was the first to construct a unital embedded in $\mathrm{Hgh}(q^2)$ for all odd prime powers $q$. His idea was to take the canonical equation of a classical unital in $PG(2, q^2)$ and show that the points of $\mathrm{Hgh}(q^2)$ satisfying that equation must form a unital in the Hughes plane, using the model described above. Kestenband [154] slightly generalized this result by showing that the absolute (Hughes) points of any Hermitian form associated with a symmetric matrix over $K = \mathrm{GF}(q)$ form a unital in $\mathrm{Hgh}(q^2)$. For a given $q$, these unitals are all equivalent to the one found by Rosati, and we refer to any unital from this equivalence class as a **classical-Rosati** unital.

Wantz [221] then constructed a class of unitals in the Hughes plane which are analogues of the orthogonal-Buekenhout-Metz unitals in $\mathrm{PG}(2, q^2)$ in the sense that the points may be described in a manner similar to that given in Section 4.2.1. Just as the classical unital is included in the class of orthogonal-Buekenhout-Metz unitals in $\mathrm{PG}(2, q^2)$, the classical-Rosati unital is included in this new class of unitals. Using left-normalized vectors to represent points in both $\mathrm{Hgh}(q^2)$ and $\mathrm{PG}(2, q^2)$, define

$$H_{ab} = \left\{ \left(1, y, ay^2 + by^{q+1} + t\epsilon\right) : y \in \mathrm{GF}(q^2), t \in \mathrm{GF}(q) \right\} \cup \{(0,0,1)\},$$

for fixed $a, b \in K = \mathrm{GF}(q)$. It is important to note that the parameters $a$ and $b$ of $H_{ab}$ are in the subfield $K = \mathrm{GF}(q)$, and also note the $\epsilon$ appearing in the definition of $H_{ab}$.

Using the coordinatization techniques developed in Section 4.2.1, one sees that $H_{ab}$ is an orthogonal-Buekenhout-Metz unital in $\mathrm{PG}(2, q^2)$ subject to certain restrictions on $a$ and $b$; namely, if and only if $b^2 - a^2$ is a nonzero square in $\mathrm{GF}(q)$. Since $(H_{ab})^q = H_{ab}$, Theorem 5.15 may be applied to obtain the following theorem of Wantz [221].

**Theorem 5.16.** *Let $q$ be an odd prime power, and let $a, b \in \mathrm{GF}(q)$ with $b^2 - a^2$ a nonzero square in $\mathrm{GF}(q)$. Then $H_{ab}$ is a unital in $\mathrm{Hgh}(q^2)$.*

We immediately see that in the case when $a = 0$ and $b \neq 0$, the points of $H_{ab}$ satisfy the Hermitian equation $xz^q + x^q z - 2by^{q+1} = 0$. Thus, $H_{0b}$ is a classical-Rosati unital for any nonzero $b \in \mathrm{GF}(q)$. In [221] the "linear" collineation group stabilizing $H_{ab}$ is computed, and equivalences are sorted out among the $H_{ab}$'s using the full automorphism group of $\mathrm{Hgh}(q^2)$. It turns out that the number of equivalence classes is precisely one less than the number of equivalence classes of orthogonal-Buekenhout-Metz unitals embedded in $\mathrm{PG}(2, q^2)$ (see Theorem 4.14). It should also be noted that it is shown in [221] that the dual unital $H_{ab}^*$ is isomorphic to $H_{a,-b}$. Since all classical-Rosati unitals ($a = 0$) are equivalent, we see that $H_{ab}$ is self-dual provided either $a = 0$ or $b = 0$. If $a \neq 0$ and $b \neq 0$, it is currently unknown if $H_{ab}$ is self-dual as a design. If $b = 0$, the unital $H_{ab}$ is known to contain at least one oval. It is presently unknown if unitals other than the $H_{ab}$'s are embedded in $\mathrm{Hgh}(q^2)$.

There are many other families of square order projective planes in which it is known that unitals are embedded. For instance, Abatangelo, Larato, and Rosati [8] have shown that the derived Hughes planes of order $q^2$ with $q \equiv 3 \pmod{4}$ always contain unitals, while Barlotti and Lunardon [31] have shown that unitals exist in the Bose-Barlotti $\Delta$-planes. In fact, the authors are unaware of any square order projective plane in which it has been shown that unitals are not embedded.

# 6

# Combinatorial Questions and Associated Configurations

In this chapter we discuss in more detail the combinatorial nature of embedded unitals. This will naturally lead to a discussion of several related combinatorial objects, as well as some interesting combinatorial questions.

## 6.1 Intersection Problems

We begin by studying various intersection sizes and patterns for unitals and Baer subplanes embedded in a square order projective plane, starting with the most general situation. That is, we consider the intersection of an arbitrary unital and a Baer subplane in a projective plane $\mathcal{P}$ of order $n^2$. Recall that every line of $\mathcal{P}$ meets an embedded unital in 1 or $n+1$ points, and similarly meets a Baer subplane in 1 or $n+1$ points. For either structure we call a line **tangent** if it meets the structure in 1 point, and **secant** if it meets the structure in $n+1$ points. The following result, which will turn out to be very useful when discussing various characterizations in the next chapter, was proved by Bruen and Hirschfeld [78] and independently by Grüning [127]. We follow the approach given in [78].

**Theorem 6.1.** *Let $\mathcal{P}$ be a projective plane of order $n^2$, and assume that $\mathcal{P}$ contains a Baer subplane $\mathcal{B}$ and a unital $U$. Let $b$ denote the number of secant lines to $\mathcal{B}$ which are tangent to $U$. Then*

$$|\mathcal{B} \cap U| = 2(n+1) - b.$$

*Proof.* Let $\mathcal{B}^*$ denote the tangent lines to $\mathcal{B}$ in $\mathcal{P}$, and let $U^*$ denote the tangent lines to $U$ in $\mathcal{P}$. We begin by computing $|\Omega|$, where

$$\Omega = \{(P, \ell) \mid P \in \mathcal{B}, \ell \in U^*, P \in \ell\}.$$

S. Barwick, G. Ebert, *Unitals in Projective Planes*,
DOI: 10.1007/978-0-387-76366-8_6, © Springer Science+Business Media, LLC 2008

Defining

$$\Omega_1 = \{(P,\ell) \mid P \in \mathcal{B} \cap U,\ \ell \in U^*,\ P \in \ell\},$$
$$\Omega_2 = \{(P,\ell) \mid P \in \mathcal{B} \backslash U,\ \ell \in U^*,\ P \in \ell\},$$
$$\Omega_3 = \{(P,\ell) \mid P \in \mathcal{B},\ \ell \in U^* \cap \mathcal{B}^*,\ P \in \ell\},$$
$$\Omega_4 = \{(P,\ell) \mid P \in \mathcal{B},\ \ell \in U^* \backslash \mathcal{B}^*,\ P \in \ell\},$$

we see that

$$|\Omega_1| + |\Omega_2| = |\Omega| = |\Omega_3| + |\Omega_4|.$$

Letting $s = |\mathcal{B} \cap U|$ and using the combinatorics of embedded unitals as discussed in Section 2.3, we compute

$$|\Omega_1| = s, \qquad\qquad |\Omega_2| = (n^2 + n + 1 - s)(n+1),$$
$$|\Omega_3| = n^3 + 1 - b, \quad |\Omega_4| = b(n+1).$$

Substituting into the above equation and solving for $s = |\mathcal{B} \cap U|$, the result follows immediately.    $\square$

The same technique may be used to prove other intersection results in this general setting. Elementary proofs of the following results may be found in [78].

**Theorem 6.2.** *Let $\mathcal{P}$ be a projective plane of order $n^2$, and assume that $\mathcal{P}$ contains a Baer subplane $\mathcal{B}$ and a unital $U$. Let $\mathcal{B}^*$ denote the set of tangent lines to $\mathcal{B}$ in $\mathcal{P}$, and let $\mathcal{B}^{**}$ denote the set of secant lines to $\mathcal{B}$. Similarly, define $U^*$ and $U^{**}$. Then*

1. $|\mathcal{B}^* \cap U^*| - |\mathcal{B} \cap U| = (n+1)(n^2 - n - 1),$
2. $|\mathcal{B}^* \cap U^{**}| + |\mathcal{B} \cap U| = n^4 - n^3 + n + 1,$
3. $|\mathcal{B}^{**} \cap U^{**}| - |\mathcal{B} \cap U| = n^2 - n - 1.$

**Theorem 6.3.** *Let $U_1$ and $U_2$ be two unitals embedded in a projective plane $\mathcal{P}$ of order $n^2$. Using the notation of Theorem 6.2, we have*

1. $|U_1 \cap U_2| = |U_1^* \cap U_2^*|,$
2. $|U_1^{**} \cap U_2^{**}| - |U_1 \cap U_2| = (n^2 - n + 1)(n^2 - n - 1).$

Theorem 6.1 immediately implies that the maximum size for the intersection of any unital and any Baer subplane in a projective plane of order $n^2$ is $2(n+1)$. In Theorem 2.9 we determined all possible intersections for a classical unital and a Baer subplane in the Desarguesian plane $\mathrm{PG}(2,q^2)$; namely, the possible intersections are a point, a Baer subline, a conic of a Baer subplane, or a pair of Baer sublines. Thus the possible intersection sizes in this classical setting are 1, $q+1$, and $2q+1$. This same pattern holds in certain cases for the intersection of an ovoidal-Buekenhout-Metz unital and a Baer subplane in $\mathrm{PG}(2,q^2)$.

**Theorem 6.4.** *Let $U$ be an ovoidal-Buekenhout-Metz unital with respect to the line $\ell_\infty$ in the Desarguesian plane $\mathrm{PG}(2, q^2)$, and let $\mathcal{B}$ be a Baer subplane in $\mathrm{PG}(2, q^2)$.*

1. *If $\mathcal{B}$ meets $\ell_\infty$ in a Baer subline, then $\mathcal{B} \cap U$ is a point, a Baer subline, an oval of a Baer subplane, or a pair of Baer sublines.*
2. *If $\mathcal{B}$ meets $\ell_\infty$ in the point $P_\infty = U \cap \ell_\infty$, then $|\mathcal{B} \cap U| = q + 1$ or $2q + 1$. Moreover, this intersection either consists of $P_\infty$ together with $q$ other points on distinct lines of $\mathcal{B}$ through $P_\infty$, or is a Baer subline through $P_\infty$ together with $q$ other points on distinct Baer sublines through $P_\infty$.*

*Proof.* Let $\mathcal{U}$ be the ovoidal cone representing $U$ in the Bruck-Bose representation for $\mathcal{P}(\mathcal{S}) = \mathrm{PG}(2, q^2)$, and let $V$ be the vertex of this cone. Suppose first that $\mathcal{B}$ meets $\ell_\infty$ in a Baer subline. Then, since we are in a Desarguesian plane $\mathrm{PG}(2, q^2)$, we know from Theorem 3.13 that in the Bruck-Bose representation $\mathcal{B}$ is represented by a plane $\pi$ of $\mathrm{PG}(4, q)$ that meets $\Sigma_\infty$ in a line $m$ which is not a spread line. In particular, $\pi$ does not contain the spread line $p_\infty$ corresponding to the point $P_\infty = U \cap \ell_\infty$. If $V \in \pi$, then $\pi \cap \mathcal{U}$ is either the point $V$, a generator of $\mathcal{U}$ other than $p_\infty$, or a pair of generators of $\mathcal{U}$ (neither of which is $p_\infty$). Hence in $\mathrm{PG}(2, q^2)$, $\mathcal{B}$ meets $U$ in a point, a Baer subline, or a pair of Baer sublines. If $V \notin \pi$, then $\pi \cap \mathcal{U}$ is a point or an oval as discussed prior to Theorem 1.19. Hence in $\mathrm{PG}(2, q^2)$, $\mathcal{B}$ meets $U$ in a point or an oval of $\mathcal{B}$, proving (1).

Now suppose that $\mathcal{B}$ meets $\ell_\infty$ in the point $P_\infty = \ell_\infty \cap U$. Then $\mathcal{B}$ is represented in $\mathrm{PG}(4, q)$ via the Bruck-Bose correspondence by a ruled cubic surface $B$ with line directrix $p_\infty$. Let $g_1, g_2, \ldots, g_{q+1}$ be the skew generators of $B$, each of which meets $p_\infty$ in one point. Let $g_1$ be the generator of $B$ which passes through the vertex $V$ of the ovoidal cone $\mathcal{U}$. For $i = 1, 2, \ldots, q + 1$, the plane $\pi_i = \langle g_i, p_\infty \rangle$ contains at least one generator of the ovoidal cone $\mathcal{U}$, namely, $p_\infty$. Moreover, by Theorem 1.19 no $\pi_i$ is a tangent plane to $\mathcal{U}$ at $p_\infty$ since no $\pi_i$ is contained in the tangent hyperplane $\Sigma_\infty$ at $p_\infty$. Thus each $\pi_i$ meets $\mathcal{U}$ in another generator $\ell_i$ of $\mathcal{U}$ (in addition to $p_\infty$). It is possible that $\ell_1 = g_1$, but certainly $\ell_i \neq g_i$ for $i \neq 1$. Thus for $i \neq 1$, in the plane $\pi_i$ the line $\ell_i$ meets $g_i$ in precisely one point. For $i = 1$, either $\ell_1 = g_1$ or $\ell_1 \cap g_1 = V$. As each $g_i$ represents a Baer subline of $\mathrm{PG}(2, q^2)$ in the Bruck-Bose representation, (2) now follows directly. □

Determining the possible intersections of a Baer subplane $\mathcal{B}$ and an ovoidal-Buekenhout-Metz unital $U$ in $\mathrm{PG}(2, q^2)$ when $\mathcal{B}$ meets $\ell_\infty$ in a point other than $P_\infty = U \cap \ell_\infty$ is much more difficult. In fact, not too much is known in this case. The following result, when $U$ is an orthogonal-Buekenhout-Metz unital, is found in the Ph.D. thesis of Quinn [182] (this result is stated in [88]). The proof, which we omit, is a lengthy discussion of various cases based on the algebraic geometry of a certain sextic curve that naturally arises.

**Theorem 6.5.** *Suppose that $U$ is an orthogonal-Buekenhout-Metz unital with respect to the line $\ell_\infty$ and $\mathcal{B}$ is a Baer subplane embedded in $\mathrm{PG}(2,q^2)$. If $q > 13$, then*

$$1 \leq |\mathcal{B} \cap U| \leq 2q + 1.$$

It should be noted that from Theorem 6.1, as discussed in a completely general situation above, we know that $0 \leq |\mathcal{B} \cap U| \leq 2q + 2$. Thus the above result simply eliminates the possible intersection sizes 0 and $2q + 2$ when the plane is Desarguesian, the unital is orthogonal-Buekenhout-Metz, and $q > 13$. Yet this minor improvement is quite difficult to prove. It should also be noted that the proof of the above result shows that as $q$ increases, more possible intersection sizes can be eliminated. An interesting consequence of Theorem 6.5 is the following.

**Corollary 6.6.** *Let $U$ be a unital in $\mathrm{PG}(2,q^2)$, where $q > 13$ is odd. If there exists a Baer subplane $\mathcal{B}$ of $\mathrm{PG}(2,q^2)$ with no point in common with $U$, then $U$ is not a Buekenhout unital.*

*Proof.* From Theorem 4.5 we know that every nonsingular Buekenhout unital embedded in $\mathrm{PG}(2,q^2)$ is classical. Moreover, as discussed in Section 4.2.1, for odd $q$ every ovoidal-Buekenhout-Metz unital in $\mathrm{PG}(2,q^2)$ is necessarily an orthogonal-Buekenhout-Metz unital. The result now follows from Theorem 6.5 and Theorem 2.9.    □

For small values of $q$, it is possible for a Buekenhout unital and a Baer subplane in $\mathrm{PG}(2,q^2)$ to share no points. For instance, a simple computer search using the software package Magma [84] finds such examples for $q = 3, 5, 7$, and 11. Extensive, but not exhaustive, computer searches found no such examples for $q = 4, 8, 9$, and 13. In fact, all possible intersection sizes between 0 and $2q + 2$ occur for $q = 3, 5, 7$, and 11. Intersection sizes found for $q = 9$ and 13 were 1 through $2q + 1$, inclusive. For $q = 4$ and 8, the intersection sizes found were all odd integers between 1 and $2q + 1$. For $q = 8$, all these odd intersection sizes were obtained when using only orthogonal-Buekenhout-Metz unitals and also when using only Buekenhout-Tits unitals.

We now discuss the intersection between two unitals in $\mathrm{PG}(2,q^2)$. This appears to be a much more difficult problem. However, in the classical setting a result of Kestenband [152] gives a complete solution. The proof given in [152] is algebraic, relying on arguments involving minimal and characteristic polynomials. We state the result here without proof.

**Theorem 6.7.** *The possible intersection patterns for two distinct classical unitals* $\mathcal{H}_1$ *and* $\mathcal{H}_2$ *in* $PG(2, q^2)$ *are the following (the configurations are illustrated in Figure 6.1):*

1. *a single point,*
2. *a Baer subline,*
3. *a* $(q^2 - q + 1)$*-arc (that is,* $q^2 - q + 1$ *points with no 3 collinear),*
4. *the union of q Baer sublines concurrent in a point* $P \in \mathcal{H}_1 \cap \mathcal{H}_2$,
5. *the union of* $q - 1$ *Baer sublines whose extended lines are concurrent in a point* $Q \notin \mathcal{H}_1 \cap \mathcal{H}_2$ *together with two other points,*
6. *the intersection of the lines in two Baer subpencils (that is, cones whose vertex is a point and whose base is a Baer subline)* $\mathcal{K}$ *and* $\mathcal{K}'$ *with vertices A and* $A'$, *respectively, such that* $A \in \mathcal{H}_1 \cap \mathcal{H}_2, A' \notin \mathcal{H}_1 \cap \mathcal{H}_2, AA' \notin \mathcal{K}$, *and* $AA' \in \mathcal{K}'$,
7. *the intersection of the lines in two Baer subpencils* $\mathcal{K}$ *and* $\mathcal{K}'$ *with vertices A and* $A'$, *respectively, such that* $A \notin \mathcal{H}_1 \cap \mathcal{H}_2, A' \notin \mathcal{H}_1 \cap \mathcal{H}_2, AA' \notin \mathcal{K}$, *and* $AA' \notin \mathcal{K}'$.

*Thus the possible intersection sizes are* $1, q + 1, q^2 - q + 1, q^2 + 1, q^2 + q + 1$, *and* $(q + 1)^2$, *with two different configurations of size* $q^2 + 1$. *Moreover, all seven intersection patterns occur.*

In [122] Giuzzi shows that any two classical unital intersections of the same type are projectively equivalent. He also computes the homography subgroup preserving each of these seven possible intersection configurations. Several authors have shown that the arc in (3) is complete (see Section 6.3.2). Donati and Durante [100, 101] study these intersections in more detail. They show that in the Bruck-Bose representation of $PG(2, q^2)$ in $PG(4, q)$, the sets (5), (6), (7) correspond to three-dimensional quadrics in some hyperplane of $PG(4, q)$. More precisely, the intersection (5) corresponds to an elliptic quadric, (6) corresponds to a conic cone, and (7) corresponds to a hyperbolic quadric. Further, in intersection (7), the set of $(q + 1)^2$ points is actually triply ruled. That is, there are three Baer subpencils which cover the points of the intersection; each Baer subpencil has a vertex not in $\mathcal{H}_1 \cap \mathcal{H}_2$. Moreover, these $3(q + 1)$ Baer sublines are the only Baer sublines contained in the intersection.

For an ovoidal-Buekenhout-Metz unital and a classical unital embedded in the Desarguesian plane $PG(2, q^2)$, the possible intersection patterns seem very hard to describe. However, it has been conjectured that the size of such an intersection is always congruent to 1 modulo $q$, just as for the intersection of two classical unitals. For instance, extensive (but not exhaustive) searching using the software package Magma [84] for small values of $q$ yields Table 6.1. Each column gives the intersection sizes found for the intersection of a classical unital and a certain family of ovoidal-Buekenhout-Metz unitals in $PG(2, q^2)$. The first column represents all the orthogonal-Buekenhout-Metz unitals which are not a union of conics (we call these "ordinary" in

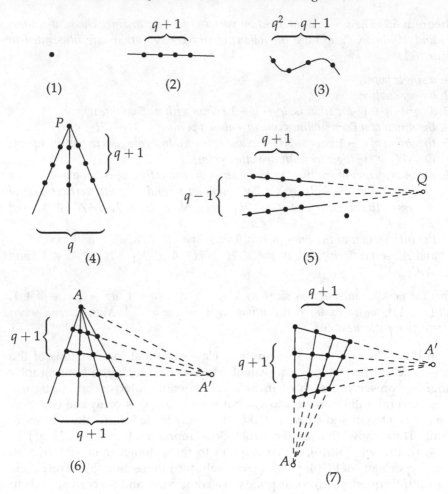

**Fig. 6.1.** Possible intersections of two classical unitals

the table), the second column represents the equivalence class of orthogonal-Buekenhout-Metz unitals which are a union of conics (necessarily for odd $q$), and the third column represents the Buekenhout-Tits unitals (for odd powers of 2).

Note that all intersection sizes in Table 6.1 are indeed congruent to 1 modulo $q$. The best known result to date in this regard is a modulo $p$ result, where $q$ is a power of the prime $p$. To discuss this result, we need to introduce some coding theory. Consider the vector space $W$ of dimension $q^4 + q^2 + 1$ over the prime field $GF(p)$, where the coordinate positions for vectors in $W$ correspond to the points of $PG(2, q^2)$ in some fixed order. If $A$ is any subset of points in $PG(2, q^2)$, then $v^A$ will denote the vector in $W$ with coordinate 1 in those positions corresponding to the points in $A$ and with coordinate 0 in all

**Table 6.1.** Intersection sizes found for the classical unital with a unital of the listed type

| $q$ | "ordinary" orthogonal-Buekenhout-Metz | union of conics | Buekenhout-Tits |
|---|---|---|---|
| 4 | 1, 9, 13, 17, 21, 25, 29, 33 | | |
| 5 | 6, 16, 21, 26, 31, 46 | 1, 6, 16, 21, 26, 31, 36, 46, 51 | |
| 7 | 1, 8, 36, 43, 50, 57, 64, 92, 99 | 8, 36, 43, 50, 57, 64, 92 | |
| 8 | 9, 49, 57, 65, 73, 81, 121, 129 | | 25, 33, 41, 49, 57, 65, 73, 81, 89, 105, 113, 121 |
| 9 | 10, 64, 73, 82, 91, 100, 154 | 10, 64, 73, 82, 91, 100, 154 | |
| 11 | 100, 111, 122, 133, 144, 232 | 100, 111, 122, 133, 144, 232 | |
| 13 | 14, 144, 157, 170, 183, 196, 326 | 1, 14, 144, 157, 170, 183, 196, 326, 339 | |
| 16 | 225, 241, 257, 273, 289, 497 | | |
| 32 | 33, 961, 993, 1025, 1057, 1089 | | 865, 897, 929, 961, 993, 1025, 1057, 1089, 1121, 1153, 1249 |

other positions; that is, $v^A$ is the **characteristic vector** of the point set $A$. Let $\mathcal{C}_p = \mathcal{C}_p(2, q^2)$ denote the subspace of $W$ spanned by the characteristic vectors of all lines in $\mathrm{PG}(2, q^2)$. This subspace $\mathcal{C}_p$ is often called the **linear code** of $\mathrm{PG}(2, q^2)$. Note that since the field of scalars for $W$ is the prime field $\mathrm{GF}(p)$, every vector in $\mathcal{C}_p$ may be expressed as a sum of characteristic vectors of certain (not necessarily distinct) lines of $\mathrm{PG}(2, q^2)$. The following result was first proved by Blokhuis, Brouwer and Wilbrink [56], although an alternate proof may be found in the book by Assmus and Key [16, Theorem 6.7.1]. In fact, this result is the response to an earlier conjecture made by Assmus and Key. We state the theorem here without proof.

**Theorem 6.8.** *Let $q = p^e$ be a prime power, and let $U$ be a unital embedded in the Desarguesian plane $\mathrm{PG}(2, q^2)$. Then $U$ is classical if and only if $v^U \in \mathcal{C}_p$, where $\mathcal{C}_p$ is the linear code over $\mathrm{GF}(p)$ spanned by the characteristic vectors of the lines in $\mathrm{PG}(2, q^2)$.*

As a corollary of this theorem, one immediately obtains the following result concerning the intersection size of a classical unital $\mathcal{H}$ and an arbitrary unital $U$ embedded in $\mathrm{PG}(2, q^2)$.

**Corollary 6.9.** *Let $\mathcal{H}$ be a classical unital and let $U$ be an arbitrary unital embedded in $\mathrm{PG}(2, q^2)$, for some prime power $q = p^e$. Then*

$$|U \cap \mathcal{H}| \equiv 1 \pmod{p}.$$

*Proof.* We work in the linear code $\mathcal{C}_p$ of $\mathrm{PG}(2, q^2)$ over the prime field $\mathrm{GF}(p)$, as described above. By Theorem 6.8, we may express

$$v^{\mathcal{H}} = v^{m_1} + v^{m_2} + \cdots + v^{m_t},$$

where $m_1, m_2, \ldots, m_t$ are (not necessarily distinct) lines of $\mathrm{PG}(2, q^2)$. Now $v^{\mathcal{H}} \cdot v^{\mathcal{H}} = |\mathcal{H}| = q^3 + 1 \equiv 1 \pmod{p}$, where $\cdot$ denotes the usual dot product. However, using the above expression for $v^{\mathcal{H}}$, we have

$$
\begin{aligned}
v^{\mathcal{H}} \cdot v^{\mathcal{H}} &= v^{\mathcal{H}} \cdot (v^{m_1} + v^{m_2} + \cdots + v^{m_t}) \\
&= (v^{\mathcal{H}} \cdot v^{m_1}) + (v^{\mathcal{H}} \cdot v^{m_2}) + \cdots + (v^{\mathcal{H}} \cdot v^{m_t}) \\
&\equiv 1 + 1 + 1 + \cdots + 1 \pmod{p} \\
&\equiv t \pmod{p},
\end{aligned}
$$

since every line of $\mathrm{PG}(2, q^2)$ meets the unital $\mathcal{H}$ in 1 or $q + 1$ points. Hence, comparing these two computations, we see that $t \equiv 1 \pmod{p}$.

Finally,

$$
\begin{aligned}
v^{U} \cdot v^{\mathcal{H}} &= v^{U} \cdot (v^{m_1} + v^{m_2} + \cdots + v^{m_t}) \\
&= (v^{U} \cdot v^{m_1}) + (v^{U} \cdot v^{m_2}) + \cdots + (v^{U} \cdot v^{m_t}) \\
&\equiv 1 + 1 + 1 + \cdots + 1 \pmod{p} \\
&\equiv t \pmod{p} \\
&\equiv 1 \pmod{p}.
\end{aligned}
$$

That is, $|U \cap \mathcal{H}| \equiv 1 \pmod{p}$, proving the result.    □

Using a completely different approach, Corollary 6.9 was proved by Baker and Ebert [21] in the case when $U$ is an orthogonal-Buekenhout-Metz unital which can be expressed as a union of conics (for odd $q$). Also it should be noted that this coding theoretic characterization of the classical unital can be adapted to other planes. In [24] it is shown that the classical-Rosati unital is the only unital, up to equivalence, among the unitals defined by Wantz in the Hughes plane (see Section 5.4.2) which lies in the linear code of that plane. Thus an analogue to the modulo $p$ result of Corollary 6.9 is valid for the intersection size of the classical-Rosati unital and any other unital embedded in $\mathrm{Hgh}(q^2)$.

To conclude this section, we note that the intersection size of two arbitrary unitals in $\mathrm{PG}(2, q^2)$ can be almost anything. In particular, this size does not have to be congruent to 1 modulo $p$. However, it should be mentioned that when $q$ is even, computations using Magma when the two unitals are ovoidal-Buekenhout-Metz have yielded only odd intersection sizes. Also, it is still unknown if disjoint unitals in $\mathrm{PG}(2, q^2)$ exist; no disjoint pairs have yet been found to the best of our knowledge. Furthermore, the intersection problem in non-Desarguesian planes is more or less wide open, except for the modulo $p$ result in the Hughes plane mentioned above.

## 6.2 Spreads and Packings

Just as one can define a (line) spread of $PG(3, q)$, one can also define a *spread* of many combinatorial objects. This certainly applies to designs in general, and unitals in particular. Recall that $t$-designs were defined in Section 2.3.

**Definition 6.10.** *Let $\mathcal{D}$ be $t$-design. Then a* **spread** *of $\mathcal{D}$ is any collection of mutually disjoint blocks that partition the points of $\mathcal{D}$. A* **packing** *of $\mathcal{D}$ is any collection of mutually disjoint spreads that partition the blocks of $\mathcal{D}$.*

It should be noted that in the context of designs, packings are often called *1-resolutions* or *parallelisms*, and the associated spreads are often called *resolution classes* or *parallel classes*. However, the words spread and packing are typically used when the design is some sort of geometry or arises naturally from some geometry. Hence, if $U$ is unital of order $n$, then a spread of $U$ is any collection of $n^2 - n + 1$ mutually disjoint blocks of $U$, which necessarily partition the $n^3 + 1$ points of $U$; and a packing of $U$ is any collection of $n^2$ mutually disjoint spreads, which necessarily partition the $n^2(n^2 - n + 1)$ blocks of $U$. Since $U$ is a 2-design with $\lambda = 1$, blocks from distinct spreads of some packing meet in at most one point.

Of course, not all unitals admit a packing. For instance, as mentioned in Section 2.3, the nonexhaustive computer search in [43] found 909 mutually nonisomorphic unitals on 28 points (that is, of order 3). The nature of the search forced all of the resulting unitals to have a spread, although 217 of them had only one spread. Moreover, only 4 of the unitals found in that search had a packing, two of which had a unique packing. In [203] another nonexhaustive computer search found 38 mutually nonisomorphic unitals on 65 points embedded in the known projective planes of order 16, fifteen of which had packings.

However, we will show that packings exist for all ovoidal-Buekenhout-Metz unitals (embedded in some translation plane). In particular, the Hermitian curve (embedded in $PG(2, q^2)$) admits several packings.

We now consider any unital $U$ of order $n$ embedded in a projective plane of order $n^2$. Let $P$ be a point of $U$, and let $\ell$ be the unique tangent line to $U$ at $P$. Using the terminology developed in [102], we say that $(P, \ell)$ is a *special flag* if the feet of every point $Q \in \ell \backslash P$ form a collinear set of points, necessarily containing the point $P$.

**Lemma 6.11.** *Let $U$ be a unital of order $n$ embedded in some projective plane $\mathcal{P}$ of order $n^2$. If $U$ has a special flag, then $U$ admits a packing.*

*Proof.* Suppose that $U$ has a special flag $(P, \ell)$. Then for each point $Q \in \ell \backslash P$, the block consisting of the feet of $Q$ together with the $n^2 - n$ blocks obtained by intersecting $U$ with the nontangent lines through $Q$ will form a spread of $U$. As $Q$ varies over the $n^2$ points of $\ell \backslash P$, the resulting $n^2$ spreads form a packing of $U$.                                                                         $\square$

We now use this notion of a special flag to prove the following result of Dover [102].

**Theorem 6.12.** *Let $U$ be an ovoidal-Buekenhout-Metz unital with respect to the line $\ell_\infty$, embedded in some translation plane $\mathcal{P}(\mathcal{S})$ of order $q^2$ which is at most two-dimensional over its kernel. If $P_\infty = U \cap \ell_\infty$, then $(P_\infty, \ell_\infty)$ is a special flag of $U$.*

*Proof.* Let $m$ be any line of $\mathcal{P}(\mathcal{S})$ through $P_\infty$, other than $\ell_\infty$. Thus $m$ is a secant line meeting $U$ in precisely $q + 1$ points, one of which is $P_\infty$. Working dually, to show that $(P_\infty, \ell_\infty)$ is a special flag, it suffices to show that the tangent lines to $U$ at the points of $m \cap U$ are concurrent at a point of $\ell_\infty$.

In the Bruck-Bose representation of the translation plane $\mathcal{P}(\mathcal{S})$, the line $\ell_\infty$ corresponds to the spread $\mathcal{S}$ of $\Sigma_\infty$. Let $\mathcal{U}$ be the ovoidal cone in $\mathrm{PG}(4, q)$ representing $U$. The line $m$ is represented in $\mathrm{PG}(4, q)$ by a plane $\pi_m$ which meets $\mathcal{U}$ in a pair of intersecting lines, one of which is the spread line $p_\infty$ corresponding to the point $P_\infty$. Let $h$ be the other generator of $\mathcal{U}$ contained in $\pi_m$, and let $\Gamma$ be the unique tangent hyperplane to $\mathcal{U}$ at $h$. Let $\ell$ be the tangent line to $U$ at some point of $m \cap U$ distinct from $P_\infty$. Thus $\ell \neq \ell_\infty$. In $\mathrm{PG}(4, q)$, $\ell$ is represented by a plane $\pi_\ell$ which meets $\mathcal{U}$ in precisely one point, necessarily a point of the generator $h$. By Theorem 1.19, this forces $\pi_\ell$ to be contained in the tangent hyperplane $\Gamma$. Moreover, from the Bruck-Bose representation we know that $\pi_\ell$ meets $\Sigma_\infty$ in a line, other than $p_\infty$, of the spread $\mathcal{S}$. Since $\Gamma$ and $\Sigma_\infty$ are distinct hyperplanes of $\mathrm{PG}(4, q)$, they must meet in a plane. A standard counting argument shows that this plane contains a unique line $t$ of the spread $\mathcal{S}$. Note that $t \neq p_\infty$. Since $\pi_\ell$ meets $\Sigma_\infty$ in a line of the spread $\mathcal{S}$, but is fully contained in the hyperplane $\Gamma$, this forces $\pi_\ell$ to contain the spread line $t$.

Since $\ell$ was an arbitrary tangent line to $U$ at a point of $m \cap U$, other than $P_\infty$, we see that every such tangent line is represented by a plane meeting $\mathcal{U}$ in one point and containing the spread line $t$. Further, the tangent line $\ell_\infty$ to $U$ at $P_\infty$ is represented by the spread $\mathcal{S}$, which also contains the line $t$. In the translation plane $\mathcal{P}(\mathcal{S})$, this means that the $q + 1$ tangent lines to $U$ at the points of $m \cap U$ are concurrent at the point $T$ of $\ell_\infty$ corresponding to the spread line $t$. Since $m$ was an arbitrary secant line through $P_\infty$, this implies by duality that $(P_\infty, \ell_\infty)$ is a special flag of $U$.    $\square$

**Corollary 6.13.** *Every ovoidal-Buekenhout-Metz unital admits a packing.*

**Corollary 6.14.** *The dual of any ovoidal-Buekenhout-Metz unital admits a packing.*

*Proof.* If $U$ is any ovoidal-Buekenhout-Metz unital, then $U$ has the special flag $(P_\infty, \ell_\infty)$. Hence the dual unital has the special flag $(\ell_\infty, P_\infty)$, and thus admits a packing by Lemma 6.11.    $\square$

Note that the classical unital $\mathcal{H} = \mathcal{H}(2, q^2)$ embedded in $\mathrm{PG}(2, q^2)$ admits at least $q^3 + 1$ packings since every (point, tangent line) flag of $\mathcal{H}$ is special; that is, the feet of any point of $PG(2, q^2) \setminus \mathcal{H}$ are collinear. Of course, these $q^3 + 1$ packings are all projectively equivalent, as are all the associated spreads. An exhaustive computer search discussed in [105] shows that $\mathcal{H}(2, 16)$ has four mutually inequivalent packings, one being given by Corollary 6.13 above. It is currently unknown how many inequivalent packings of $\mathcal{H}(2, q^2)$ exist for $q > 4$.

We now concentrate on spreads of the classical unital $\mathcal{H} = \mathcal{H}(2, q^2)$. In addition to the spreads described above, there are many others in this classical setting. For instance, exhaustive computer searches in [106] show that there are 3 mutually inequivalent spreads of $\mathcal{H}(2, 16)$, 10 mutually inequivalent spreads of $\mathcal{H}(2, 25)$, and 81 mutually inequivalent spreads of $\mathcal{H}(2, 49)$. All spreads of $\mathcal{H}(2, 9)$ are projectively equivalent. It thus appears that as $q$ increases, the number of mutually inequivalent spreads of $\mathcal{H}(2, q^2)$ grows quite rapidly. The asymptotic nature of this growth is currently unknown.

In [23] a spread of $\mathcal{H}(2, q^2)$ is called **orthogonally divergent** if no spread line (that is, a line containing a block of the spread as a Baer subline) contains the pole of any spread line, including itself. Such spreads are of importance because of their connection to certain problems in coding theory. It is shown in [23] that orthogonally divergent spreads of $\mathcal{H}(2, q^2)$ exist when $q$ is even and $q \equiv 1 \pmod 3$. Moreover, the orthogonally divergent spread constructed in this case admits a cyclic group of order $q^2 - q + 1$ acting sharply transitively on its blocks, and hence the spread is called *cyclic*. It is further shown in [23] that cyclic spreads of $\mathcal{H}(2, q^2)$ exist if and only if $q$ is even, and cyclic orthogonally divergent spreads of $\mathcal{H}(2, q^2)$ exist if and only if $q$ is even and $q \equiv 1 \pmod 3$.

Most of the spreads of $\mathcal{H}(2, q^2)$ found in the previously mentioned computer searches for $q = 4, 5$, and 7 arise in the following way. This idea first appeared in [23] in a slightly different context. Let $\mathcal{H} = \mathcal{H}(2, q^2)$ be the nondegenerate Hermitian curve (classical unital) in $PG(2, q^2)$ defined by $X_0^{q+1} + X_1^{q+1} + X_2^{q+1} = 0$, as described in Section 2.1. Then the points $P = (1, 0, 0)$, $Q = (0, 1, 0)$, and $R = (0, 0, 1)$ of $PG(2, q^2)$ are the vertices of a self-polar triangle with respect to $\mathcal{H}$. That is, each side of this triangle is a secant line of $\mathcal{H}$ and the opposite vertex of the triangle is the pole of that secant line with respect to the associated unitary polarity. Then the $q^3 + 1$ points of $\mathcal{H}$ are partitioned by the chords on the sides of the triangle $\triangle PQR$ and the sets

$$T_a = \{(1, y, z) \mid y^{q+1} = -(1 + a), \ z^{q+1} = a\},$$

as $a$ varies over $GF(q) \setminus \{0, -1\}$. Each $T_a$ has $(q + 1)^2$ points of $\mathcal{H}$ by our discussion of norms in Section 1.2. Moreover, each $T_a$ is triply-ruled, in the sense that there are three sets of $q + 1$ secant lines to $\mathcal{H}$, one such set through each of $P$, $Q$, and $R$, whose intersections with $\mathcal{H}$ precisely cover $T_a$. Using dual coordinates, for a specific $a$ these sets of secant lines are

$$H_a = \{(\alpha, 0, 1)' \mid \alpha^{q+1} = a\},$$
$$V_a = \{(\nu, 1, 0)' \mid \nu^{q+1} = -(1+a)\},$$
$$D_a = \{(0, \delta, 1)' \mid \delta^{q+1} = -a/(1+a)\}.$$

Hence one can create a spread of $\mathcal{H}$ by selecting one of the three ruling families ($H_a$, $V_a$, or $D_a$) of $T_a$ for each value of $a$ and then adjoining the chords on the three sides of the triangle $\triangle PQR$. Thus the sets $T_a$ can be thought of as *switching sets*, and one can create new spreads of $\mathcal{H}$ from old ones by replacing one ruling family on $T_a$ by another ruling family on $T_a$ for various choices of $a$. If one chooses the ruling family $H_a$ for all choices of $a$, then the spread obtained is projectively equivalent to one of the spreads described in the proof of Lemma 6.11 above. Similarly, no new spreads are obtained if one chooses the ruling family $V_a$ for all choices of $a$, or if one chooses the ruling family $D_a$ for all choices of $a$. However, mixed choices for the ruling families as $a$ varies will yield a spread that is not equivalent to one of those described in the proof of Lemma 6.11. This is completely analogous to constructing the André spreads from a regular spread of $PG(3, q)$, as discussed at the end of Section 3.4.3, since the switching sets used above are "linear" in a well-defined sense.

In [105] Dover generalizes the above process by showing how "nonlinear" switching sets may be used to obtain even more spreads of $\mathcal{H}$. The method of Dover seems to be quite robust, as all but 2 of the 81 inequivalent spreads of $\mathcal{H}(2, 49)$ may be obtained in this way. Similarly, all but 2 of the 10 inequivalent spreads of $\mathcal{H}(2, 25)$ can be obtained from Dover's construction. It would be of interest to see if these four spreads are sporadic or if they are part of some other infinite family (or families) of spreads that can be obtained by some other construction.

Finally, to complete this section we mention spreads and packings of the Ree unital. Recall from Section 2.3 that the Ree unital of order $q$, where $q$ is an odd power of 3, cannot be embedded in any projective plane in such a way that the Ree group carries over. In fact, there is no known embedding of the Ree unital of order $q$ in any projective plane of order $q^2$. Thus the techniques described above for constructing spreads and packings cannot be applied to this unital. Nonetheless, it turns out that the Ree unital has at least $q^3 + 1$ packings, one naturally associated with each point of the unital. These packings are all equivalent, and any two of them share precisely one spread (see [104]). This is the same situation as that described after the proof of Corollary 6.14 for the classical unital, although the spreads and packings are constructed in very different ways. The constructions in the case of the Ree unital are completely group-theoretic.

## 6.3 Related Combinatorial Structures

In this section we discuss some combinatorial structures that are naturally related to unitals.

### 6.3.1 Inversive Planes

Previously the only designs we considered were 2-designs. In this section we look at a certain family of 3-designs.

**Definition 6.15.** *Let* $n \geq 2$ *be an integer. An* **inversive plane** *of* **order** $n$ *is any* $3$-$(n^2 + 1, n + 1, 1)$ *design.*

Since three distinct points of an inversive plane uniquely determine a block, the blocks are often called **circles**. In fact, an inversive plane is a type of *circle geometry*, which alternately can be defined in a natural axiomatic way, although we will not do so here. One way to construct inversive planes is the following. Let $\mathcal{O}$ be an ovoid in $PG(3, q)$. Thus $\mathcal{O}$ consists of $q^2 + 1$ points in $PG(3, q)$, no three of which are collinear. Moreover, every plane in $PG(3, q)$ meets $\mathcal{O}$ in a single point or an oval. A plane which meets $\mathcal{O}$ in one point is called a **tangent plane** of $\mathcal{O}$, while a plane meeting $\mathcal{O}$ in an oval is called a **secant plane** of $\mathcal{O}$. Since any three distinct points of $\mathcal{O}$ determine a unique plane, which is necessarily a secant plane, we can construct an inversive plane of order $q$ by taking the points of $\mathcal{O}$ as our set of points and the secant plane intersections with $\mathcal{O}$ as our set of blocks. Such inversive planes are called **egg-like**, and all known finite inversive planes are egg-like. The inversive plane obtained from an elliptic quadric in $PG(3, q)$ is called **Miquelian** because it satisfies a certain configurational result of Miquel. It is often denoted by $M(q)$. Similarly, the Suzuki-Tits inversive plane, obtained from the Suzuki-Tits ovoid, when $q \geq 8$ is an odd power of 2, is often denoted by $S(q)$. It should be noted that the Miquelian inversive plane, $M(q)$, alternatively may be obtained by taking the points of $PG(1, q^2)$ as the set of points and the Baer sublines of $PG(1, q^2)$ as the set of blocks.

For a thorough discussion of inversive planes the interested reader should see Chapter 6 of Dembowski [96]. Here we mention only a few important facts. Straightforward counting shows that in an inversive plane of order $n$, there are $n(n + 1)$ blocks through any point $P$. Similarly, given any two distinct points $P$ and $Q$ in an inversive plane of order $n$, there are $n + 1$ blocks through these two points, and any point $R$ other than $P$ or $Q$ lies on exactly one of these blocks. For a given point $P$, one may define a new incidence structure by taking the points other than $P$ as the new point set and the blocks through $P$, with $P$ deleted from each such block, as the new block set. Incidence is that induced from the original inversive plane. Then one immediately sees that this new incidence structure is a 2-$(n^2, n, 1)$ design; that is, an affine plane of order $n$. This affine plane is called the **internal structure** at the point $P$. For an egg-like inversive plane obtained from an ovoid in $PG(3, q)$, the internal structure at any point $P$ is isomorphic to the Desarguesian affine plane $AG(2, q)$. The **point-residual** of an inversive plane is the incidence structure obtained by deleting a point $P$ and all the blocks incident with $P$. From our computations above we see that that the point-residual of an inversive plane of order $n$ is a 2-$(n^2, n + 1, n)$ design.

We now show that with every ovoidal-Buekenhout-Metz unital there is a naturally associated point-residual of an egg-like inversive plane. We follow the general approach given by Dover [102]. This result was proved independently by Barwick and O'Keefe [37]. In the special case of orthogonal-Buekenhout-Metz unitals embedded in the Desarguesian plane $PG(2, q^2)$, this result was proved by Baker and Ebert [22] ($q$ odd) and Ebert [108] ($q$ even).

**Theorem 6.16.** *Let $U$ be an ovoidal-Buekenhout-Metz unital with respect to the line $\ell_\infty$, embedded in some translation plane $\mathcal{P}(\mathcal{S})$ of order $q^2$ which is at most two-dimensional over its kernel. Let $\mathcal{U}$ be the ovoidal cone in $PG(4, q)$ representing $U$ in the Bruck-Bose representation for $\mathcal{P}(\mathcal{S})$. Let $P_\infty = U \cap \ell_\infty$, and consider an incidence structure $\mathcal{D}$ whose points are the chords of $U$ passing through $P_\infty$. For each chord $m$ of $U$ not through $P_\infty$, define $B_m$ to be the set of chords of $U$ passing through $P_\infty$ which meet $m$ in some point, and take as the blocks of $\mathcal{D}$ the distinct sets $B_m$ as $m$ varies. Then $\mathcal{D}$ is isomorphic to the point-residual of the egg-like inversive plane associated with any ovoidal hyperplane section of $\mathcal{U}$.*

*Proof.* Since $\mathcal{U}$ is an ovoidal cone in $PG(4, q)$, any hyperplane $\Sigma$ not passing through the vertex $V$ of $\mathcal{U}$ meets $\mathcal{U}$ in an ovoid $\mathcal{O}$. Now $\mathcal{O}$ meets $\Sigma_\infty$ in a point $Q$ on the spread line $p_\infty$ corresponding to $P_\infty = U \cap \ell_\infty$.

Consider the design $\mathcal{D}'$, whose points are the points of $\mathcal{O} \setminus Q$, whose blocks are the intersections of $\mathcal{O}$ with the secant planes of $\Sigma$ not passing through $Q$, and whose incidence is given by containment. This design is clearly the point-residual of the egg-like inversive plane determined by the base ovoid $\mathcal{O}$. We now establish an isomorphism between the incidence structure $\mathcal{D}$ in the statement of the theorem and this design $\mathcal{D}'$.

By the Bruck-Bose correspondence, the $q^2$ chords of $U$ through $P_\infty$ correspond to the $q^2$ generators of $\mathcal{U}$ other than $p_\infty$. If $b$ is a chord of $U$ through $P_\infty$, we also let $b$ denote the corresponding generator of $\mathcal{U}$. Thus we define a mapping $\theta$ from the point set of $\mathcal{D}$ to the point set of $\mathcal{D}'$ via the rule that maps the chord $b$ to the point $b \cap \mathcal{O}$. Clearly, $\theta$ is a bijection between the points of $\mathcal{D}$ and the points of $\mathcal{D}'$.

We next show that $\theta$ maps blocks of the incidence structure $\mathcal{D}$ to blocks of the design $\mathcal{D}'$. To this end, let $m$ be a chord of $U$ which does not contain the point $P_\infty$. Then $B_m$ is the block of $\mathcal{D}$ consisting of the $q + 1$ chords of $U$ through $P_\infty$ that meet $m$. Let $\pi_m$ be the plane of $PG(4, q)$ which represents the line of $\mathcal{P}(\mathcal{S})$ containing the chord $m$. For each $b \in B_m$, the plane $\pi_m$ meets the line $b$ in one point. Thus $\pi_m$ must meet $\mathcal{U}$ in an oval, and the $q + 1$ lines $b \in B_m$ must form a three-dimensional oval cone $\mathcal{C}$ in $PG(4, q)$.

Hence the image of the block $B_m$ under $\theta$ is the intersection of the oval cone $\mathcal{C}$ with the ovoid $\mathcal{O}$. Since $\mathcal{O}$ lies in the hyperplane $\Sigma$ and $\mathcal{C}$ lies in some hyperplane $\Gamma$, distinct from $\Sigma$, we have $\mathcal{C} \cap \mathcal{O} = (\Gamma \cap \Sigma) \cap \mathcal{U}$. Now $\Gamma \cap \Sigma$ is a plane that meets $\mathcal{U}$ in an oval since each of the $q + 1$ lines of $\mathcal{C}$ meets $\Sigma$ in a point. Moreover, this oval does not contain the point $Q$ as all of its points are affine. Therefore, the image of the block $B_m$ is an oval of $\mathcal{O}$ which does not contain the point $Q$; that is, the image of any block in $\mathcal{D}$ under $\theta$ is a block of the design $\mathcal{D}'$.

Finally, it remains to show that every block of $\mathcal{D}'$ is the image of some block $B_m$. To do this, let $\mathcal{A}$ be any oval of $\mathcal{O}$ which does not contain $Q$. If $\pi$ is the plane of $\Sigma$ meeting $\mathcal{O}$ in $\mathcal{A}$, then $\Delta = \langle V, \pi \rangle$ is a hyperplane in $PG(4, q)$ meeting $\mathcal{U}$ in the oval cone with vertex $V$ and base $\mathcal{A}$. Clearly, $\Delta \neq \Sigma_\infty$ and thus $\Delta \cap \Sigma_\infty$ is a plane, which necessarily contains a unique line $\ell$ of the spread $\mathcal{S}$. Since $Q \notin \mathcal{A}$, this plane does not contain $p_\infty$ and thus $\ell \neq p_\infty$. Now, there exists a plane $\pi'$ through $\ell$ which meets the oval cone $\Delta \cap \mathcal{U}$ in an oval, say $\mathcal{B}$. Thus $\mathcal{B}$ and $\mathcal{A}$ are oval sections of the same oval cone. Under the Bruck-Bose correspondence, let $m'$ be the line of $\mathcal{P}(\mathcal{S})$ represented by the plane $\pi'$. By definition the mapping $\theta$ will send the block $B_{m'}$ of $\mathcal{D}$ to the block $\mathcal{A}$ of $\mathcal{D}'$, and $\theta$ is an isomorphism between $\mathcal{D}$ and $\mathcal{D}'$. In particular, the incidence structure $\mathcal{D}$ is a $2\text{-}(q^2, q + 1, q)$ design which is isomorphic to the point-residual of the egg-like inversive plane obtained from the ovoid $\mathcal{O}$ in $\Sigma$.                                                                    $\square$

We note that in Section 7.3.3, we briefly discuss some results due to Wilbrink [223] who shows that if an arbitrary unital satisfies two special geometric conditions, then it gives rise to an inversive plane.

### 6.3.2 Arcs

We now return to one of the results from the previous section; namely, the possible intersection patterns for two classical unitals in $PG(2, q^2)$. Recall that one such intersection pattern is a set of $q^2 - q + 1$ points in $PG(2, q^2)$, no three of which are collinear. Such a set of points is called an *arc*.

**Definition 6.17.** *A* **k-arc** *is a set of $k$ points in $PG(2, q)$, no three of which are collinear. If a $k$-arc is not contained in any $(k + 1)$-arc, then the $k$-arc is called* **complete**.

The largest size for a $k$-arc in $PG(2, q)$ is $q + 1$ for odd $q$ and $q + 2$ for even $q$ (see [60]). The $(q + 1)$-arcs in $PG(2, q)$ are the ovals introduced in Section 1.4. Typically, $(q + 2)$-arcs in $PG(2, q)$ are called **hyperovals**. It should be noted that these maximum sizes for a $k$-arc remain valid in non-Desarguesian planes (see [184]). A problem that historically has received a considerable amount of attention is the problem of determining the largest size for a complete arc in $PG(2, q)$ which is not an oval or hyperoval; that is, the problem of determining the size of the second largest complete arc in $PG(2, q)$. A result of Segre [199], using techniques from algebraic geometry, shows that this second largest size for even $q$ is at most $q - \sqrt{q} + 1$. In fact, this bound can be achieved for all even powers of 2 that are greater than 4. Namely, in Theorem 6.7 one of the possible intersections for two distinct Hermitian curves in $PG(2, q^2)$ is an arc of size $q^2 - q + 1$. It is shown, independently, by Boros and Szőnyi in [58] and by Fisher, Hirschfeld, and Thas in [118] that these arcs are complete for $q > 2$. Hence the above Segre bound is met for all even powers of 2 that are greater than 4. For odd $q$, these complete arcs are not necessarily of the second largest size.

It is interesting to note that the above $(q^2 - q + 1)$-arcs in $PG(2, q^2)$ are point orbits under a certain cyclic group that naturally acts on the Desarguesian plane. Moreover, one can express the Hermitian curve (classical unital) as a disjoint union of $q + 1$ such arcs. To do this, it is most convenient to use a *field model* for the Desarguesian plane. That is, we consider the finite field $GF(q^6)$ as a three-dimensional vector space over its subfield $GF(q^2)$, and use this vector space to model $PG(2, q^2)$. Hence if $\alpha$ is a nonzero field element in $GF(q^6)$, we let $(\alpha)$ denote the corresponding projective point in $PG(2, q^2)$. Note that in this setting, the subfield $GF(q^3)$, which is a three-dimensional vector space over its subfield $GF(q)$, represents a Baer subplane of $PG(2, q^2)$. This is true since in the lattice of subfields for $GF(q^6)$, we have $GF(q^3) \cap GF(q^2) = GF(q)$. Of course, there are other subspaces of $GF(q^6)$ which also represent the same Baer subplane.

To coordinatize this situation, we let $\beta$ denote a primitive element of $GF(q^6)$, so that $\zeta = \beta^{q^4 + q^2 + 1}$ is a primitive element of $GF(q^2)$. Then the points of $PG(2, q^2)$ are $(\beta^0), (\beta^1), \ldots, (\beta^{q^4 + q^2})$. Multiplication by $\beta$ thus induces a collineation (in fact, a homography) $\theta$ acting on $PG(2, q^2)$, which is often called a **Singer cycle**. The cyclic group $S$ generated by $\theta$ has order $q^4 + q^2 + 1$ and is called the **Singer group** acting on $PG(2, q^2)$. The Singer group has a sharply transitive action on both points and lines of $PG(2, q^2)$. Choosing the (cyclic) subgroup of $S$ of order $q^2 + q + 1$ and looking at the point orbits under this subgroup, the partition $\mathcal{B}(\theta) = \{\mathcal{B}_0, \mathcal{B}_1, \ldots \mathcal{B}_{q^2-q}\}$ is obtained, where

$$\mathcal{B}_i = \left\{ \left( \beta^{i + s(q^2 - q + 1)} \right) \mid s = 0, 1, \ldots, q^2 + q \right\}$$

is a Baer subplane for each $i = 0, 1, \ldots, q^2 - q$. Indeed, $\mathcal{B}_0$ is the Baer subplane represented by $GF(q^3)$ and $\mathcal{B}(\theta)$ is its orbit under $S$.

Choosing the subgroup of $S$ of order $q^2 - q + 1$, the orbits $\mathcal{A}(\theta) = \{\mathcal{A}_0, \mathcal{A}_1, \ldots, \mathcal{A}_{q^2+q}\}$, where

$$\mathcal{A}_j = \left\{ \left( \beta^{j + t(q^2 + q + 1)} \right) \mid t = 0, 1, \ldots, q^2 - q \right\}$$

for $j = 0, 1, \ldots, q^2 + q$, partition the points of $PG(2, q^2)$ into $q^2 + q + 1$ subsets of size $q^2 - q + 1$. We now show that each $\mathcal{A}_j$ is a $(q^2 - q + 1)$-arc.

**Lemma 6.18.** *Using the above notation, each $\mathcal{A}_j$ is an arc.*

*Proof.* To simplify the notation, we let $N$ denote the integer $q^2 + q + 1$. Since $\mathcal{A}(\theta)$ is an orbit of $\mathcal{A}_0$ under the Singer group $S$, it suffices to show that $\mathcal{A}_0$ is an arc. By way of contradiction, suppose that three points of $\mathcal{A}_0$ are collinear. Since $\mathcal{A}_0$ is a point orbit under a subgroup of $S$, we may assume without loss of generality that the three collinear points are $(1)$, $(\beta^{iN})$, and $(\beta^{jN})$, where $1 \leq i < j \leq q^2 - q$. These points are collinear if and only if $1 + \alpha\beta^{iN} = \delta\beta^{jN}$, for some nonzero elements $\alpha, \delta \in \mathrm{GF}(q^2)$.

Now $(1 + \alpha\beta^{iN})^{q^3+1} = (\delta\beta^{jN})^{q^3+1}$, where $(\delta\beta^{jN})^{q^3+1} = \delta^{q+1}\beta^{jN(q^3+1)}$ is in $\mathrm{GF}(q)$ and

$$(1 + \alpha\beta^{iN})^{q^3+1} = (1 + \alpha\beta^{iN})(1 + \alpha\beta^{iN})^{q^3}$$
$$= 1 + \alpha\beta^{iN} + \alpha^q\beta^{iNq^3} + \alpha^{q+1}\beta^{iN(q^3+1)}.$$

Since $\alpha^{q+1}\beta^{iN(q^3+1)} \in \mathrm{GF}(q)$, we have $\alpha\beta^{iN} + \alpha^q\beta^{iNq^3} \in \mathrm{GF}(q)$.

Hence $f(x) = (x - \alpha\beta^{iN})(x - \alpha^q\beta^{iNq^3})$ is a quadratic polynomial over $\mathrm{GF}(q)$ having $\alpha\beta^{iN}$ as a root. Thus $\alpha\beta^{iN}$ is in $\mathrm{GF}(q^2)$, and hence $\beta^{iN} \in \mathrm{GF}(q^2)$. This implies that $i$ is a multiple of $q^2 - q + 1$, a contradiction. Hence each $\mathcal{A}_j$ is an arc of size $q^2 - q + 1$.    $\square$

Thus $\mathcal{B}(\theta)$ is a partition of $\mathrm{PG}(2, q^2)$ into Baer subplanes, and $\mathcal{A}(\theta)$ is a partition of $\mathrm{PG}(2, q^2)$ into $(q^2 - q + 1)$-arcs. Moreover, $|\mathcal{A}_i \cap \mathcal{B}_i| = 1$ for all $i$ and $j$, since $\gcd(q^2 - q + 1, q^2 + q + 1) = 1$.

For the remainder of this section, we primarily follow the development in [58]. Hence, in order to introduce Hermitian curves into the picture, we let $T$ denote the trace from $\mathrm{GF}(q^6)$ to its subfield $\mathrm{GF}(q^2)$. That is,

$$T : x \mapsto x + x^{q^2} + x^{q^4}.$$

Straightforward computations now show that the mapping

$$s : \mathrm{GF}(q^6) \times \mathrm{GF}(q^6) \to \mathrm{GF}(q^2)$$
$$(x, y) \mapsto T(xy^{q^3})$$

is a nondegenerate Hermitian form. The associated unitary polarity has the nondegenerate Hermitian curve

$$\mathcal{H}_1 = \left\{ (x) \mid T(x^{q^3+1}) = 0 \right\}$$

in $\mathrm{PG}(2, q^2)$ as its set of absolute points in $\mathrm{PG}(2, q^2)$. In fact, one similarly sees that

$$\mathcal{H}_\alpha = \left\{ (x) \mid T(\alpha x^{q^3+1}) = 0 \right\}$$

is a nondegenerate Hermitian curve for each nonzero $\alpha$ in the subfield $\mathrm{GF}(q^3)$.

**Lemma 6.19.** *Using the above notation, if the arc $\mathcal{A}_j$ has at least one point on the Hermitian curve $\mathcal{H}_\alpha$, then $\mathcal{A}_j$ is contained in $\mathcal{H}_\alpha$.*

*Proof.* Let $(x)$ be a point of $\mathcal{A}_j \cap \mathcal{H}_\alpha$, so that $T(\alpha x^{q^3+1}) = 0$. Any other point of $\mathcal{A}_j$ looks like $y = x\beta^{i(q^2+q+1)}$, for some $i = 1, 2, \ldots, q^2 - q$. Now

$$
\begin{aligned}
T(\alpha y^{q^3+1}) &= T(\alpha x^{q^3+1} \beta^{i(q^2+q+1)(q^3+1)}) \\
&= T(\beta^{i(q^4+q^2+1)(q+1)} \alpha x^{q^3+1}) \\
&= \beta^{i(q^4+q^2+1)(q+1)} T(\alpha x^{q^3+1}) \\
&= 0,
\end{aligned}
$$

since $\beta^{i(q^4+q^2+1)(q+1)} \in \mathrm{GF}(q) \subset \mathrm{GF}(q^2)$. Hence $y \in \mathcal{H}_\alpha$ and the result follows. $\square$

**Theorem 6.20.** *Every nondegenerate Hermitian curve in $\mathrm{PG}(2, q^2)$ can be partitioned into cyclic $(q^2 - q + 1)$-arcs.*

*Proof.* As any two nondegenerate Hermitian curves are projectively equivalent, we may work with the Hermitian curve $\mathcal{H}_1$, as defined above. Since $\mathcal{A}(\theta)$ is a partition of $\mathrm{PG}(2, q^2)$ into cyclic $(q^2 - q + 1)$-arcs, the result follows from Lemma 6.19. $\square$

**Theorem 6.21.** *For odd $q$, every nondegenerate Hermitian curve (classical unital) in $\mathrm{PG}(2, q^2)$ can be partitioned into Baer subconics.*

*Proof.* As any two nondegenerate Hermitian curves are projectively equivalent, we may work with the Hermitian curve $\mathcal{H}_1$, as defined above. Consider the Baer subplane $\mathcal{B}_0$, represented by the subfield $\mathrm{GF}(q^3)$. The trace function $T$, when restricted to $\mathrm{GF}(q^3)$, has the rule $T(x) = x + x^q + x^{q^2}$ since $x^{q^4} = x^q$ for $x \in \mathrm{GF}(q^3)$. Thus $T$ restricted to $\mathrm{GF}(q^3)$ is the trace $T_0$ from $\mathrm{GF}(q^3)$ to $\mathrm{GF}(q)$. Hence the previously defined sesquilinear form $s$ when so restricted becomes $s_0 : \mathrm{GF}(q^3) \times \mathrm{GF}(q^3) \mapsto \mathrm{GF}(q)$ via $s_0(x, y) = T_0(xy)$. Thus $s_0$ is a nondegenerate symmetric bilinear form that induces an orthogonal polarity on the Baer subplane $\mathcal{B}_0$. As discussed in Section 1.5, this polarity is ordinary since $q$ is odd. Hence the absolute points in $\mathcal{B}_0$ form a Baer subconic. This Baer subconic is the intersection $\mathcal{H}_1 \cap \mathcal{B}_0$.

Since $\mathcal{H}_1$ is the disjoint union of the $(q^2 - q + 1)$-arcs which each pass through some point of $\mathcal{H}_1 \cap \mathcal{B}_0$, the result now follows by applying the Singer subgroup of order $q^2 - q + 1$ to this Baer subconic. $\square$

It should be noted that if $q$ is even, then the orthogonal polarity on the Baer subplane $\mathcal{B}_0$ induced by $s_0$ will be a pseudopolarity, and hence the absolute points will form a Baer subline (alternately, see the discussion prior to Theorem 2.10). Thus the argument in the above proof shows that for even $q$ the Hermitian curve can be partitioned into Baer sublines. Of course, we already knew this could be done for any $q$ since in Section 6.2 we showed that Hermitian curves admit spreads, and each chord of a Hermitian curve is a Baer subline. However, this argument does show that the Hermitian curve admits cyclic spreads for even $q$ (see the discussion concerning orthogonally divergent spreads in Section 6.2).

We now return to the Hermitian curves $\mathcal{H}_\alpha$, as $\alpha$ varies over the nonzero elements of $GF(q^3)$. Since $GF(q^3) \cap GF(q^2) = GF(q)$, we see from the form of the associated unitary polarities that $\mathcal{H}_{\alpha_1} = \mathcal{H}_{\alpha_2}$ if and only if $\alpha_1/\alpha_2 \in GF(q)$. Thus we obtain $q^2 + q + 1$ distinct Hermitian curves in this way, each one being the disjoint union of $q + 1$ arcs from the arc partition $\mathcal{A}(\theta)$. In fact, any two of these Hermitian curves meet precisely in one arc from $\mathcal{A}(\theta)$.

**Theorem 6.22.** *Let $\alpha_1$ and $\alpha_2$ be nonzero elements of $GF(q^3)$ whose ratio is not in the subfield $GF(q)$. Then the Hermitian curves $\mathcal{H}_{\alpha_1}$ and $\mathcal{H}_{\alpha_2}$ meet precisely in one of the $(q^2 - q + 1)$-arcs from $\mathcal{A}(\theta)$.*

*Proof.* Since $\alpha_1/\alpha_2 \notin GF(q)$, the Hermitian arcs $\mathcal{H}_{\alpha_1}$ and $\mathcal{H}_{\alpha_2}$ are distinct. From Theorem 6.7 we know that these two Hermitian curves cannot be disjoint. If $(x)$ is any point in their intersection, then the $(q^2 - q + 1)$-arc of $\mathcal{A}(\theta)$ passing through $(x)$ must lie on both of these Hermitian curves. From the list of possible intersections given in Theorem 6.7, we see that $\mathcal{H}_{\alpha_1} \cap \mathcal{H}_{\alpha_2}$ is precisely this arc. $\square$

Thus we see that the cyclic $(q^2 - q + 1)$-arcs in $\mathcal{A}(\theta)$ are precisely the (complete) arcs of size $q^2 - q + 1$ obtained as the intersection of certain pairs of the above nondegenerate Hermitian curves, all such arcs being projectively equivalent. Straightforward counting shows that there are precisely $q + 1$ Hermitian curves $\mathcal{H}_\alpha$ containing a given arc in $\mathcal{A}(\theta)$, and any such arc is the intersection of any two of these $q + 1$ Hermitian curves. Thus we have another model for a projective plane of order $q$; namely, take as "points" the arcs in $\mathcal{A}(\theta)$, take as "lines" the distinct Hermitian curves $\mathcal{H}_\alpha$, and let incidence be given by set inclusion.

As a final comment on these arcs, we mention that it is shown in [118] that the tangents of any of the above complete $(q^2 - q + 1)$-arcs in $PG(2, q^2)$ form a dual classical unital if and only if $q$ is even.

An interesting question is whether the known unitals in $PG(2, q^2)$ contain larger arcs. This question has been answered for ovals in [14, 22, 108, 133]. The classical unital and the Buekenhout-Tits unitals do not contain ovals. The only orthogonal-Buekénhout-Metz unitals that contain an oval are the special class of orthogonal-Buekenhout-Metz unitals that are a pencil of conics. The proofs of these results appear in Section 4.2.

## 6.4 Unitals and Codes

One of the most interesting questions posed concerning codes and unitals was the conjecture made by Assmus and Key that was mentioned in

Section 6.1. Let $\mathrm{PG}(2, q^2)$ be the Desarguesian projective plane of square order $q^2$, where $q = p^e$ for some prime $p$. Recall that $\mathcal{C}_p = \mathcal{C}_p(2, q^2)$ denotes the linear code over the prime field $\mathrm{GF}(p)$ generated by the characteristic vectors of the lines of $\mathrm{PG}(2, q^2)$. The conjecture made was that among all the unitals embedded in $\mathrm{PG}(2, q^2)$, only the classical unital (nondegenerate Hermitian curve) has its characteristic vector in the code $\mathcal{C}_p$. As previously mentioned, this conjecture was proved to be true in 1991 by Blokhuis, Brouwer, and Wilbrink [56], and this implies that any unital embedded in $\mathrm{PG}(2, q^2)$ meets the classical unital in 1 (mod $p$) points.

In general, a **linear code** $\mathcal{C}$ is any subspace of a finite-dimensional vector space defined over some field. The **dimension** of the code is the dimension of this subspace $\mathcal{C}$, and the **length** of the code is the number of coordinate positions. The linear code is called **binary** if the field of scalars is $\mathrm{GF}(2)$. The **weight** of a vector is the number of nonzero components in that vector, and the **minimum weight** of a linear code is the least weight among all nonzero vectors in the code. The minimum weight of a linear code determines how many errors the code can correct, and the dimension of the code indicates how many distinct codewords can be sent. As a final piece of notation, we let $\mathcal{C}^\perp$ denote the orthogonal complement of the subspace $\mathcal{C}$ with respect to the usual dot product. For a good introduction to coding theory, see [47] or [129].

The codes we consider in this section are slightly different than the codes $\mathcal{C}_p(2, q^2)$ discussed in Section 6.1. Namely, we consider the unital $U$ as a $2\text{-}(q^3 + 1, q + 1, 1)$ design, independent of whether it is embedded in any projective plane of order $q^2$, and look at the code of length $q^3 + 1$ generated by the characteristic vectors of its blocks over some prime field $\mathrm{GF}(r)$, where $r$ is a prime number and the positions are identified with the points of the unital in some fixed order. We denote this linear code by $\mathcal{C}_r(U)$.

The only known unitary designs for which the dimension of the code $\mathcal{C}_r(U)$ has been completely determined are the Ree unitals (see [135]) and the classical unitals (see [136]). In both cases, an explicit formula for the dimension of the code was determined by Hiss, using characters and modular representation theory. We only state the formula for the classical unital, where

$$\dim(\mathcal{C}_r(\mathcal{H}(2, q^2))) = \begin{cases} q(q^2 - q + 1) & \text{if } r \mid (q + 1) \\ q^3 + 1 & \text{otherwise.} \end{cases}$$

Partial results and special cases of this result had previously been proved by a number of other researchers, as this problem had attracted considerable attention before finally being solved completely.

As discussed in Section 4.1, the only known unitals embedded in the Desarguesian plane $\mathrm{PG}(2, q^2)$ are the ovoidal-Buekenhout-Metz unitals, one of which is the classical unital. We restrict to odd prime powers $q$, and consider the binary codes of the nonclassical orthogonal-Buekenhout-Metz unitals. As shown in Section 4.2.1, for each odd $q$ there is a unique, up to projective equivalence, nonclassical orthogonal-Buekenhout-Metz unital which can be expressed as a union of $q$ conics in $\mathrm{PG}(2, q^2)$, pairwise tangent at the point $P_\infty$. It is the only orthogonal-Buekenhout-Metz unital, including the classical unital, which contains a conic, and the only conics contained in this unital are the $q$ conics mentioned above. Let $U_0$ denote this "special" orthogonal-Buekenhout-Metz unital, and let $C_i$, for $i = 1, 2, \ldots, q$, denote the $q$ conics which pairwise meet in $P_\infty$ and whose union is $U_0$. Let $\mathcal{C}_2(U_0)$ denote the binary linear code spanned by the characteristic vectors of the blocks of $U_0$. For any set $S$ of points in $U_0$, let $v^S$ denote its characteristic vector, as previously defined.

**Theorem 6.23.** *Using the above notation, $\{v^{C_i} \mid i = 1, 2, \ldots, q\}$ is a linearly independent set of vectors in $\mathcal{C}_2(U_0)^\perp$.*

*Proof.* Using the ideas in the proof of Corollary 6.9, a vector $v$ lies in the orthogonal complement of the binary code of a design if and only if each block of the design meets the support of $v$ in an even number of points, where the support of $v$ is the set of points corresponding to the positions where $v$ has a 1. Since $U_0$ is embedded in $\mathrm{PG}(2, q^2)$, the unique tangent line to $U_0$ at a point $P$ is the unique tangent line to the conic $C_i$ containing $P$, including the case where $P = P_\infty$ and the line $\ell_\infty$ is the common tangent line to each conic $C_i$ at $P_\infty$. Thus no tangent line to any conic $C_i$ is a block of the unital $U_0$. Hence every block of $U_0$ must meet each conic in 0 or 2 points, and $v^{C_i} \in \mathcal{C}_2(U_0)^\perp$ for $i = 1, 2, \ldots, q$. As the $q$ conics are mutually tangent at $P_\infty$, the geometry implies their characteristic vectors are linearly independent over $\mathrm{GF}(2)$. $\square$

**Corollary 6.24.** *We have $\dim\left(\mathcal{C}_2(U_0)^\perp\right) \geq q$ and so $\dim(\mathcal{C}_2(U_0)) \leq q^3 - q + 1$.*

To show that equality holds in the above two inequalities, it would suffice to show that the characteristic vectors of the $q$ conics comprising $U_0$ are a spanning set for $\mathcal{C}_2(U_0)^\perp$. This can be expressed very simply from a set-theoretic point of view. Namely, let $S$ be a subset of points in $U_0$ such that $S$ meets every block of $U_0$ in an even number of points. One must then show that $S$ is a union of some subset of the $C_i$'s, where $P_\infty$ is deleted if the union is taken over an even number of $C_i$'s. Unfortunately, there does not seem to be an elementary proof of this fact.

Alternatively, to show that equality exists in the above corollary, one could exhibit a linearly independent set of $q^3 - q + 1$ vectors in $\mathcal{C}_2(U_0)$. We now describe a linearly independent set of $q^3 - q + 1$ vectors in $V(q^3 + 1, 2)$ (the ambient space of $\mathcal{C}_2(U_0)$) whose span contains $\mathcal{C}_2(U_0)$, although it is not clear that all these vectors are in $\mathcal{C}_2(U_0)$. Fix a block $b_0$ of $U_0$ which passes through $P_\infty$, and let $P_i$ be the point of $b_0 \cap C_i$, other than $P_\infty$, for $i = 1, 2, \ldots, q$. Let $Q_{ij}$, for $j = 1, 2, \ldots, q^2 - 1$, denote the points on $C_i \setminus \{P_i, P_\infty\}$, for $i = 1, 2, \ldots, q$. Note that

$$S = \left\{ v^{\{P_i, Q_{ij}\}} \mid i = 1, 2, \ldots, q; \; j = 1, 2, \ldots, q^2 - 1 \right\}$$

is a set of $q(q^2 - 1)$ weight-two vectors which are linearly independent over GF(2). As no vector in $S$ has $P_\infty$ in its support, $S \cup \{v^{b_0}\}$ is a linearly independent set of $q^3 - q + 1$ vectors in $V(q^3 + 1, 2)$. It is clear that any point pair from $C_i \setminus P_\infty$ can be realized as the support of the sum of two vectors in $S$. Since the characteristic vector of every block of $U_0$ not containing $P_\infty$ is the sum of $(q + 1)/2$ weight-two vectors of this type, and every block through $P_\infty$ (other than $b_0$) can be written as the sum of $v^{b_0}$ with $q$ carefully chosen vectors in $S$, the span of $S \cup \{v^{b_0}\}$ contains $C_2(U_0)$. If one could show that the above collection of weight-two vectors are indeed in $C_2(U_0)$, equality in the above corollary would follow immediately. Unfortunately, once again there does not seem to be an elementary proof of this fact.

However, based on computer computations for "small" values of $q$, it appears that the suggestions made above are indeed true. Moreover, in practice the weight-two vectors in $C_2(U_0)$ are precisely those vectors whose support sets $\{P, Q\}$ are contained in $C_i \setminus P_\infty$ for some $i$. The following unpublished conjecture is due to Baker and Wantz, who also provided the above suggestions for a potential elementary proof.

Conjecture 6.25 (Baker and Wantz). If $U_0$ is the special orthogonal-Buekenhout-Metz unital defined above, then $\dim(C_2(U_0)) = q^3 - q + 1$, the minimum distance of $C_2(U_0)$ is two, and the number of vectors of weight two in $C_2(U_0)$ is precisely $q\binom{q^2}{2}$.

It seems likely, based on the experience for the linear codes of the Ree and classical unitals, that representation theory will need to be used to prove this conjecture. The best partial result to date is the following lower bound of Leung and Xiang [165]:

Theorem 6.26. *Using the above notation,*

$$\dim(C_2(U_0)) \geq q^3 \left(1 - \frac{1}{p}\right) + \frac{q^2}{p}.$$

As shown in Section 4.2.1, the number of inequivalent orthogonal-Buekenhout-Metz unitals for a given odd prime power $q$ approaches infinity as $q \to \infty$. Computer computations using the software package Magma [84] indicate that all weight-two vectors are in $C_2(U)$, when $U$ is an orthogonal-Buekenhout-Metz unital which is neither classical nor the "special" union-of-conics unital. In such cases the binary code spanned is the subspace of all even weight vectors, and hence of codimension two.

Conjecture 6.27 (Baker and Wantz). If $q$ is an odd prime power and $U$ is any nonclassical orthogonal-Buekenhout-Metz unital in $PG(2, q^2)$, other than $U_0$, then $\dim(C_2(U)) = q^3$.

As a final comment on unitals and codes, we mention that Goppa [124] showed how to construct linear codes from various curves defined over finite fields. His approach is quite different from the approach described here.

In [214] it is shown that the codes so constructed from nondegenerate Hermitian curves are usually better, from a practical coding theoretic point of view, than the corresponding Reed-Solomon codes.

# Characterization Results

In this penultimate chapter we tackle the difficult problem of characterizing unitals. That is, given a set of points, what conditions on this set force it to be the point set of unital? These conditions might be geometrical, combinatorial, group theoretic, and so on. Most often, we will assume that the set of points is embedded in the Desarguesian plane $PG(2, q^2)$. We concentrate on the combinatorial and geometric characterizations here, leaving the group theoretic characterizations to a brief appendix.

## 7.1 Characterizations of Unitals via Baer Sublines

In the first section we investigate the interesting question of how many Baer sublines a unital in $PG(2, q^2)$ needs to have among its chords in order to characterize it as classical or ovoidal-Buekenhout-Metz.

We showed in Section 2.2 that every line secant to the classical unital $U$ in $PG(2, q^2)$ meets $U$ in a Baer subline. Lefèvre-Percsy [164] and Faina and Korchmáros [116] independently proved the converse; that is, if every secant line of a unital in $PG(2, q^2)$ meets it in a Baer subline, then the unital is classical. This is Theorem 7.4.

Ovoidal-Buekenhout-Metz unitals also contain many Baer sublines. If $U$ is ovoidal-Buekenhout-Metz with respect to the point $T$, then every secant through $T$ meets $U$ in a Baer subline (since lines of $PG(4, q) \setminus \Sigma_\infty$ correspond to Baer sublines of $PG(2, q^2)$). Casse, O'Keefe, and Penttila [87] and Quinn and Casse [183] proved that the converse of this is also true; see Theorem 7.6.

In order to present an overall picture of the results in this area, we state the results first, and then present the proofs in Section 7.2. The proofs of the characterizations of the classical unital are closely tied to characterizations of ovoidal-Buekenhout-Metz unitals. We present the characterizations in a convenient order with regard to the proofs given here, rather than in order of significance or in the order in which they were originally proved. The majority of these results are proved in the Bruck-Bose setting. We note that

S. Barwick, G. Ebert, *Unitals in Projective Planes*,
DOI: 10.1007/978-0-387-76366-8_7, © Springer Science+Business Media, LLC 2008

the converse of all the results stated in this section are true; that is, all these characterizations are "if and only if".

In order to prove the characterization that all chords of a unital being Baer sublines implies that the unital is classical, we use two other characterizations due to Lefèvre-Percsy. The first is a characterization of ovoidal-Buekenhout-Metz unitals; this result is also true in any translation plane of dimension at most two over its kernel.

**Theorem 7.1 ([163]).** *Let $U$ be a unital tangent to $\ell_\infty$ in $\mathrm{PG}(2, q^2)$, $q > 2$. If all Baer sublines having a point on $\ell_\infty$ intersect $U$ in 0, 1, 2, or $q + 1$ points, then $U$ is an ovoidal-Buekenhout-Metz unital with respect to $\ell_\infty$.*

The second is a characterization of the classical unital among the class of orthogonal-Buekenhout-Metz unitals.

**Theorem 7.2 ([164]).** *Let $U$ be an orthogonal-Buekenhout-Metz unital with respect to $P_\infty = U \cap \ell_\infty$ in $\mathrm{PG}(2, q^2)$, $q > 2$. If there is a secant line not through $P_\infty$ that meets $U$ in a Baer subline, then $U$ is classical.*

Theorem 7.2 was generalized to ovoidal-Buekenhout-Metz unitals by Barwick and Quinn.

**Theorem 7.3 ([39]).** *Let $U$ be an ovoidal-Buekenhout-Metz unital with respect to $P_\infty = U \cap \ell_\infty$ in $\mathrm{PG}(2, q^2)$, $q > 2$. If there is a secant line not through $P_\infty$ that meets $U$ in a Baer subline, then $U$ is classical.*

We now have enough preliminaries to show that all chords being Baer sublines implies that the unital is classical.

**Theorem 7.4 ([116, 164]).** *Let $U$ be a unital in $\mathrm{PG}(2, q^2)$, $q > 2$. If each line secant to $U$ intersects $U$ in a Baer subline, then $U$ is a classical unital.*

Lefèvre-Percsy showed that an equivalent formulation of this result is the following:
*If every Baer subline meets $U$ in 0, 1, 2, or $q + 1$ points, then $U$ is classical.*

Barwick generalized Lefèvre-Percsy's characterization of ovoidal-Buekenhout-Metz unitals (Theorem 7.1) to the nonsingular-Buekenhout unital case. As all nonsingular-Buekenhout unitals in $\mathrm{PG}(2, q^2)$ are classical (Theorem 4.5), this gives the characterization of the classical unital stated below. We note that the proof of this theorem actually gives the more general characterization of nonsingular-Buekenhout unitals in any translation plane of dimension at most two over its kernel.

**Theorem 7.5 ([33]).** *Let $U$ be a unital in $\mathrm{PG}(2, q^2)$ with $\ell_\infty$ secant to $U$. If every Baer subline with a point on $\ell_\infty$ meets $U$ in 0, 1, 2, or $q + 1$ points, then $U$ is classical.*

Recalling the characterization of ovoidal-Buekenhout-Metz unitals given in Theorem 7.1, the following improvement was given by Casse, O'Keefe, and Penttila for $q$ even, and by Quinn and Casse for all $q$.

**Theorem 7.6 ([87, 183]).** *Let $U$ be a unital in $PG(2, q^2)$, $q > 2$. If there exists a point $T$ of $U$ such that every secant line through $T$ meets $U$ in a Baer subline, then $U$ is ovoidal-Buekenhout-Metz with respect to $T$.*

This result combined with Theorem 7.3 immediately gives the following characterization of the classical unital.

**Theorem 7.7 ([39]).** *Let $U$ be a unital in $PG(2, q^2)$. If there is a point $T$ such that every secant line through $T$ meets $U$ in a Baer subline and there is a secant line not through $T$ which meets $U$ in a Baer subline, then $U$ is classical.*

That is, $q^2 + 1$ Baer sublines in this specified configuration characterize the unital as classical. To date, this is the smallest number of Baer sublines which characterize the unital as classical.

We also have the following corollary that a nonclassical Buekenhout unital is an ovoidal-Buekenhout-Metz unital with respect to exactly one tangent line.

**Corollary 7.8 ([39]).** *If $U$ is a unital in $PG(2, q^2)$ that is an ovoidal-Buekenhout-Metz unital with respect to two distinct tangent lines, then $U$ is classical.*

To complete this section we state three related results which we will not prove in this book. The first of these is a different type of characterization of the classical unital via Baer sublines. This was proved by Ball, Blokhuis, and O'Keefe using polynomial techniques.

**Theorem 7.9 ([26]).** *Let $U$ be a unital in $PG(2, p^2)$, where $p$ is a prime. If $U$ has at least $(p^2 - 2)\sqrt{p}$ Baer sublines among its chords, then $U$ is classical.*

Building upon the characterizations of unitals via Baer sublines, one can obtain characterizations of unitals by studying their intersection with Baer subplanes. In particular, Barwick, O'Keefe, and Storme obtained the following characterizations. These results also have a generalization to translation planes of dimension at most two over their kernel. The proofs use geometric techniques in the Bruck-Bose representation.

**Theorem 7.10 ([38]).** *The following statements hold:*

1. *A unital in $PG(2, q^2)$ is classical if and only if every Baer subplane meets the unital in either $1$, $q + 1$, or $2q + 1$ points.*
2. *A unital in $PG(2, q^2)$ tangent to $\ell_\infty$ is an ovoidal-Buekenhout-Metz unital with respect to $\ell_\infty$ if and only if every Baer subplane secant to $\ell_\infty$ meets $U$ in $1$, $q + 1$, or $2q + 1$ points.*

Finally, for completion, we mention a characterization of ovoidal-Buekenhout-Metz unitals given by Larato. Note that the characterization given in Theorem 7.3 is much stronger than this one.

**Theorem 7.11 ([160]).** *Let $U$ be a unital tangent to $\ell_\infty$ in $PG(2, q^2)$, $q$ odd. Then $U$ is an ovoidal-Buekenhout-Metz unital if and only if for any two affine lines $\ell$ and $m$ through $U \cap \ell_\infty$, there is an affine Baer subplane $\mathcal{B}$ such that $\mathcal{B} \cap U = (\ell \cap U) \cup (m \cap U)$.*

## 7.2 Proofs of Results from Section 7.1

The proofs presented in this section all use the Bruck-Bose representation. In particular, we use the notation of Section 3.4. Recall that $PG(2, q^2)$ with line $\ell_\infty$ at infinity is represented by $PG(4, q)$ with hyperplane $\Sigma_\infty$ at infinity containing a regular spread $S$. The points of $\ell_\infty$ correspond to the lines of $S$. If $T$ is a point of $\ell_\infty$, we denote the corresponding spread element by $t$. If $\ell$ is a line of $PG(2, q^2)$ distinct from $\ell_\infty$, we denote the corresponding plane of $PG(4, q) \backslash \Sigma_\infty$ by $\pi_\ell$. If $U$ is a unital in $PG(2, q^2)$, we denote the corresponding set of points in $PG(4, q)$ by $\mathcal{U}$.

As several results in this section rely on characterizations of sets of type $(0, 1, 2, q + 1)$, we begin by quoting these characterizations.

### Sets of type $(0, 1, 2, q + 1)$

A set of points in $PG(n, q)$ is called a **set of type $(0, 1, 2, q + 1)$** if every line of $PG(n, q)$ meets the set in $0, 1, 2$, or $q + 1$ points. Characterizations of sets of type $(0, 1, 2, q + 1)$ in projective spaces of dimension two and three were given by Lefèvre-Percsy and for "large" sets in dimension $n \geq 3$ by Tallini. The planar characterization can also be found in [130, Lemma 16.2.5].

**Theorem 7.12 ([162]).** *A set of type $(0, 1, 2, q + 1)$ in $PG(2, q)$, $q > 2$, is one of the following: an arc, a line, two lines, a nonincident point-line pair, or all of $PG(2, q)$.*

We will need a characterization of sets in dimensions 3 and 4 of type $(0, 1, 2, q + 1)$ that do not contain a plane. We start with the characterization stated in [134, Theorem 22.11.3], and specialize to the case we need. We use the notation for subspaces and quadrics given in [134]: namely, $\Pi_r$ represents a subspace of dimension $r$; $\mathcal{P}_n$ is a nonsingular parabolic quadric in a projective space of dimension $n$; and $\mathcal{H}_n$ is a nonsingular hyperbolic quadric in a projective space of dimension $n$. The notation $\Pi_r \mathcal{Q}$ refers to the cone with base $\mathcal{Q}$ and vertex $\Pi_r$.

**Theorem 7.13 ([205]).** *Let $T$ be a set of type $(0, 1, 2, q + 1)$ in $PG(n, q)$, $q > 2$, such that $T$ does not contain a plane.*

1. *If $n = 3$ and $|T| \geq q^2 + q + 1$, then $T$ is one of the following: $\mathcal{H}_3$; $\Pi_0 \mathcal{P}_2$; $\Pi_0 \mathcal{P}_2 \cup \Pi_r$, where $q$ is even and $\Pi_r$ is a subspace of the nuclear line $\Pi_0 N$ ($N$ being the nucleus of $\mathcal{P}_2$); or $\Pi_0 \mathcal{K} \cup \Pi_r$, where $q$ is even, $\mathcal{K}$ is a $(q + 1)$-arc, and $\Pi_r$ is a subspace of the nuclear line $\Pi_0 N$ ($N$ being the nucleus of $\mathcal{K}$).*
2. *If $n = 4$ and $|T| \geq q^3 + q^2 + q + 1$, then $T$ is either $\mathcal{P}_4$ or $\mathcal{P}_4 \cup N$, where in the latter case $q$ is even and $N$ is the nucleus of $\mathcal{P}_4$ (the point which is the intersection of all the tangent hyperplanes).*

**Proof of Theorem 7.1**

Let $U$ be a unital in $PG(2, q^2)$ and let $\ell_\infty$ be a tangent line of $U$. Suppose that every Baer subline with a point on $\ell_\infty$ meets $U$ in $0, 1, 2$ or $q + 1$ points. We must show that $U$ is an ovoidal-Buekenhout-Metz unital with respect to $\ell_\infty$. The proof given here follows that given by Lefèvre-Percsy [163].

We work using the Bruck-Bose representation. Let $\mathcal{U}$ denote the set of points in $PG(4, q)$ corresponding to the unital $U$. We show that $\mathcal{U}$ is an ovoidal cone meeting $\Sigma_\infty$ in a spread line by using a series of lemmas. Note that as $U \cap \ell_\infty$ is a point $T$, necessarily $\mathcal{U} \cap \Sigma_\infty$ is a line $t$ of the spread $S$.

**Lemma 7.14.** *Using the above notation, $\mathcal{U}$ is a set of type $(0, 1, 2, q + 1)$ in $PG(4, q)$.*

*Proof.* Let $P, Q$ be two points of $\mathcal{U}$. We must show that the line $PQ$ either meets $\mathcal{U}$ in exactly $P$ and $Q$, or is contained in $\mathcal{U}$. If $P, Q$ are both in $\Sigma_\infty$, then they lie on the line $t$ which is contained in $\mathcal{U}$. If at least one of $P, Q$ is not in $\Sigma_\infty$, then $PQ$ is a line of $PG(4, q)$ that meets $\Sigma_\infty$ in a point. Hence $PQ$ corresponds to a Baer subline of $PG(2, q^2)$ with a point on $\ell_\infty$. So by the hypothesis of the theorem, $PQ$ meets $U$ in $2$ or $q + 1$ points. Thus in $PG(4, q)$, the line $PQ$ meets $\mathcal{U}$ in $2$ or $q + 1$ points. $\square$

**Lemma 7.15.** *Let $\pi$ be a plane of $PG(4, q) \setminus \Sigma_\infty$ through the line $t$. Then $\mathcal{U} \cap \pi$ consists of $t$ and a further line $m$.*

*Proof.* The plane $\pi$ corresponds to a line of $PG(2, q^2)$, and this line must be secant to $U$ since the unique tangent to $U$ through $T$ is $\ell_\infty$. Hence $\pi$ meets $\mathcal{U} \setminus \Sigma_\infty$ in $q$ points. By Lemma 7.14, $\mathcal{U}$ is a set of type $(0, 1, 2, q + 1)$ and hence so is $\pi \cap \mathcal{U}$. Hence by Theorem 7.12, $\pi \cap \mathcal{U}$ is the union of two lines. $\square$

An immediate corollary of this lemma is that $\mathcal{U}$ contains no plane of $PG(4, q)$.

**Lemma 7.16.** *The set of points $\mathcal{U}$ is the union of lines through a point $V$ on $t$.*

*Proof.* By Lemma 7.15, $\mathcal{U}$ is the union of lines that meet $t$. Suppose that two of these lines, say $m$ and $m'$, meet $t$ in different points $V$ and $V'$, respectively. By Lemma 7.15, $m$ and $m'$ are not coplanar, and hence they span a hyperplane $\Sigma$ of $PG(4, q)$. Thus $\mathcal{U} \cap \Sigma$ is a set of type $(0, 1, 2, q + 1)$ that contains the skew lines $m$ and $m'$. By Lemma 7.15, every plane through $t$ in $\Sigma$ meets $\mathcal{U} \cap \Sigma$ in $t$ and one further line, except for the plane $\Sigma_\infty \cap \Sigma$, so that $|\mathcal{U} \cap \Sigma| = q^2 + q + 1$. This contradicts the classification of sets of type $(0, 1, 2, q + 1)$ in $PG(3, q)$, $q > 2$, given in Theorem 7.13. $\square$

**Lemma 7.17.** *Let $\Sigma$ be a hyperplane of $PG(4, q)$ not through $V$. Then $\mathcal{U} \cap \Sigma$ is an ovoid that meets $\Sigma_\infty$ in one point.*

*Proof.* By Lemma 7.14, $\mathcal{O} = \mathcal{U} \cap \Sigma$ is a set of type $(0, 1, 2, q+1)$. As $V \notin \Sigma$, by Lemma 7.16, $\mathcal{O}$ contains no line. Hence $\mathcal{O}$ is a cap in $\Sigma$; that is, no three points of $\mathcal{O}$ are collinear. Now $\mathcal{U}$ contains $q^3$ affine points and one point on $\ell_\infty$, hence $|\mathcal{U}| = q^3 + q + 1$. Thus $\mathcal{U}$ is the union of $q^2 + 1$ lines through $V$. Hence $|\mathcal{O}| = q^2 + 1$ and so $\mathcal{O}$ is an ovoid. Furthermore, $\mathcal{O}$ meets $\Sigma_\infty$ precisely in the point $\Sigma \cap t$.                                                                    □

Hence $\mathcal{U}$ is a cone with vertex $V$ on the spread line $t$ in $\Sigma_\infty$ and base an ovoid that meets $\Sigma_\infty$ in a point of $t$. That is, $\mathcal{U}$ is an ovoidal-Buekenhout-Metz unital. This completes the proof of Theorem 7.1. We note that this proof generalizes to give the same characterization in any translation plane of order $q^2$ with kernel containing $GF(q)$.

The converse of Theorem 7.1 is also true since in $PG(4, q)$ a line meets an ovoidal cone in 0, 1, 2, or $q+1$ points. Hence every Baer subline of $PG(2, q^2)$ with a point on $\ell_\infty$ meets an ovoidal-Buekenhout-Metz unital in 0, 1, 2, or $q+1$ points.

**Proof of Theorem 7.2**

We follow the proof given in [164], which uses coordinates in the Bruck-Bose setting. We use the coordinate notation of Section 3.4.4. Recall that points in $PG(4, q)$ have coordinates $(X_0, X_1, X_2, X_3, X_4) = (x_0, x_1, y_0, y_1, z)$ with $x_0, x_1, y_0, y_1, z \in GF(q)$, and $\Sigma_\infty$ is the hyperplane with equation $z = 0$. Points in $PG(2, q^2)$ have coordinates $(X_0, X_1, X_2) = (x, y, z)$ with $x, y, z \in GF(q^2)$, and $\ell_\infty$ has equation $z = 0$. We uniquely represent an element $\alpha \in GF(q^2)$ as $\alpha = a_0 + a_1 \zeta$, where $\zeta$ is a primitive element of $GF(q^2)$ satisfying the polynomial $x^2 - t_1 x - t_0 = 0$, and $a_0, a_1 \in GF(q)$.

Let $U$ be an orthogonal-Buekenhout-Metz unital with respect to the point $P_\infty = U \cap \ell_\infty$ in $PG(2, q^2)$, $q > 2$, and suppose that a secant line $\ell$ not through $P_\infty$ meets $U$ in a Baer subline. Then $U$ corresponds to an elliptic cone $\mathcal{U}$ in $PG(4, q)$ which meets $\Sigma_\infty$ in the spread line $p_\infty$. Let $V \in p_\infty$ be the vertex of the elliptic cone $\mathcal{U}$, and let $Q$ be another point on $p_\infty$. The secant line $\ell$ in $PG(2, q^2)$ corresponds to a plane $\pi$ of $PG(4, q) \setminus \Sigma_\infty$ containing a spread line, say $m_\pi$, distinct from $p_\infty$. Hence $\langle \pi, Q \rangle$ is a hyperplane that meets $\mathcal{U}$ in an elliptic quadric $\mathcal{E}$. There are two tangent planes to $\mathcal{E}$ through the spread element $m_\pi$, one being the plane $\langle m_\pi, Q \rangle$ contained in $\Sigma_\infty$. Let $\pi'$ be the other tangent plane to $\mathcal{E}$, and let $O = \pi' \cap \mathcal{E}$ (see Figure 7.1).

We can choose coordinates for our configuration in $PG(4, q)$ so that $O = (0, 0, 0, 0, 1)$, $V = (0, 0, 0, 1, 0)$, $Q = (0, 0, 1, 0, 0)$, $\pi$ is given by the equations $y_1 = 0$ and $y_0 = z$, and $\pi'$ is given by the equations $y_0 = y_1 = 0$. Hence in $PG(2, q^2)$ we have the following coordinates: $O = (0, 0, 1)$, $P_\infty = (0, 1, 0)$, $\ell$ has equation $y = z$, and $\ell'$ has equation $y = 0$ (see Figure 7.1).

Now by assumption, $\ell \cap U$ is a Baer subline and hence is a set of affine points $(x, 1)$ satisfying an affine equation of the form $axx^q + \alpha x + \alpha^q x^q = 1$, where $\alpha \in GF(q^2)$ and $a \in GF(q)$. Using the techniques of Section 3.4.4, we

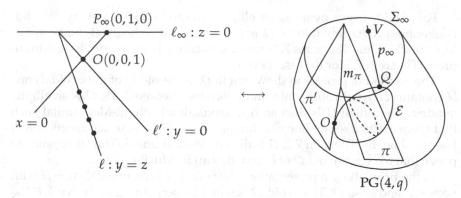

**Fig. 7.1.** Proof of Theorem 7.2

turn this into an equation in $PG(4, q)$. The corresponding affine equation has form

$$a(x_0^2 + t_1 x_0 x_1 - t_0 x_1^2) + b x_0 + c x_1 = 1,$$

where $a, b, c \in GF(q)$. This is the equation of the conic $\pi \cap \mathcal{U}$.

There is at most one elliptic quadric in $\langle \pi, Q \rangle$ that contains the conic $\pi \cap \mathcal{U}$, is tangent to $\pi'$ at $O$, and is tangent to the plane $\langle \pi, Q \rangle \cap \Sigma_\infty$ at $Q$. Note that the 3-space $\langle \pi, Q \rangle$ has equation $y_1 = 0$. Now consider the quadric $\mathcal{E}$ in $\langle \pi, Q \rangle$ which has affine equation $a(x_0^2 + t_1 x_0 x_1 - t_0 x_1^2) - y_0 = 0$. Straightforward calculations show that $\mathcal{E}$ is indeed an elliptic quadric in $\langle \pi, Q \rangle$ satisfying the above conditions, and hence is the unique such elliptic quadric. (Note that the existence of such a quadric implies that $b = c = 0$.)

Now the elliptic cone $\mathcal{U}$ in $PG(4, q)$ with vertex $V = (0, 0, 0, 1, 0)$ and base $\mathcal{E}$ also has equation $a(x_0^2 + t_1 x_0 x_1 - t_0 x_1^2) - y_0 = 0$. More straightforward computations, using the techniques of Section 3.4.4, show that the Hermitian curve in $PG(2, q^2)$ satisfying the affine equation

$$a x x^q - (2 + t_1 t_0^{-1} \zeta)^{-1} (y + (t_1 t_0^{-1} \zeta + 1) y^q) = 0$$

corresponds precisely to this elliptic cone $\mathcal{U}$ in the Bruck-Bose representation. Thus the unital $U$ is classical, proving Theorem 7.2.

**Proof of Theorem 7.3**

Let $U$ be an ovoidal-Buekenhout-Metz unital with respect to $P_\infty = U \cap \ell_\infty$ in $PG(2, q^2)$, $q > 2$. Let $\ell$ be a secant line not through $P_\infty$ that meets $U$ in a Baer subline. We follow the proof of [39], which uses the following powerful characterization of ovoids due to Brown [68].

**Theorem 7.18.** *Let $\mathcal{O}$ be an ovoid of $PG(3, q)$, $q$ even, and let $\pi$ be a plane of $PG(3, q)$ such that $\pi \cap \mathcal{O}$ is a conic. Then $\mathcal{O}$ is an elliptic quadric.*

For $q$ odd, every ovoid is an elliptic quadric ([28]), so every ovoidal-Buekenhout-Metz unital in $PG(2, q^2)$, $q$ odd, is an orthogonal-Buekenhout-Metz unital. Thus Theorems 7.3 and 7.2 coincide when $q$ is odd. It remains to prove Theorem 7.3 for the case $q$ even.

The method of proof is to show that the base ovoid $\mathcal{O}$ of the ovoidal cone $\mathcal{U}$ contains an irreducible conic, and hence by Theorem 7.18, $\mathcal{O}$ is an elliptic quadric. The unital $U$ is then an orthogonal-Buekenhout-Metz unital such that there exists a secant line not through $P_\infty = U \cap \ell_\infty$ which meets $U$ in a Baer subline. By Theorem 7.2, $U$ is then a classical unital. Thus it remains to prove that the base ovoid $\mathcal{O}$ of $\mathcal{U}$ contains an irreducible conic.

In the Bruck-Bose representation, the unital $U$ is an ovoidal cone $\mathcal{U}$ with vertex $V$ and base $\mathcal{O}$. The ovoid $\mathcal{O}$ lies in a hyperplane $\Sigma$ such that $\Sigma \cap \Sigma_\infty$ is a tangent plane $\beta$ to $\mathcal{O}$ at a point $Q \in \mathcal{O}$. Moreover, $p_\infty \not\subseteq \beta$, $Q = p_\infty \cap \beta$, and $Q \neq V$. Note that the plane $\beta$ in $\Sigma_\infty$ necessarily contains a unique line of the spread $\mathcal{S}$; denote this line by $m_\beta$.

Recall that $\ell$ is a secant line of $U$, with $P_\infty \notin \ell$, such that $\ell \cap U$ is a Baer subline $b$. Since $P_\infty$ is the unique point of $U$ on $\ell_\infty$ and since $P_\infty \notin b$, the Baer subline $b$ is disjoint from $\ell_\infty$. Thus in the Bruck-Bose representation, $b$ corresponds to an irreducible conic $b^*$ in the plane $\pi$ corresponding to $\ell$. Denote by $m_\pi$ the unique spread element in the plane $\pi$. Note that $V \notin \pi$ and $\pi \not\subseteq \Sigma_\infty$, so that $\Sigma' = \langle V, \pi \rangle$ is a hyperplane distinct from the hyperplanes $\Sigma$ and $\Sigma_\infty$ (see Figure 7.2).

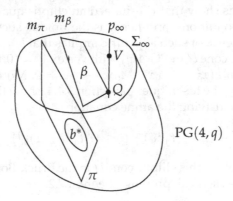

**Fig. 7.2.** Proof of Theorem 7.3

If $m_\pi = m_\beta$, then it is possible that the plane $\pi$ is contained in the hyperplane $\Sigma$; in this case $\pi \cap \mathcal{U} = \pi \cap \mathcal{O} = b^*$ and thus $\mathcal{O}$ contains the irreducible conic $b^*$.

In the remaining case, let $\mathcal{C}$ be the projection of $b^*$ from $V$ onto the hyperplane $\Sigma$. By construction of $\mathcal{U}$, $\mathcal{C}$ is a subset of $q + 1$ points of the ovoid $\mathcal{O}$. Moreover, $\mathcal{C}$ is contained in the plane $\Sigma' \cap \Sigma$, so that $\mathcal{C}$ is a $(q + 1)$-arc in a plane section of the ovoid $\mathcal{O}$.

In the hyperplane $\Sigma' = \langle V, \pi \rangle$, the point $V$ and the irreducible conic $b^*$ define a quadratic cone $\mathcal{Q}$ with vertex $V$ and base $b^*$. In fact, $\mathcal{C}$ is the intersection of $\mathcal{Q}$ with the plane $\Sigma' \cap \Sigma$, so that $\mathcal{C}$ is an irreducible conic (see [130, Section 15.3]). Thus $\mathcal{O}$ contains an irreducible conic as required.

## Proof of Theorem 7.4

Faina and Korchmáros [116] prove this theorem by taking a group theoretical approach to show that the unital satisfies the "reciprocity property" (see Definition 7.35), and hence the unital is classical by the characterization of Tallini Scafati (see Theorem 7.37). On the other hand, Lefèvre-Percsy [164] proves the theorem in the Bruck-Bose setting, and this is the proof we present here.

We begin with an equivalence condition that allows us to restate Theorem 7.4 in such a way that we may use our previous characterizations.

**Lemma 7.19.** *Let $U$ be a unital in $\mathrm{PG}(2, q^2)$. Then the following two conditions are equivalent:*

1. *every secant line of $U$ meets $U$ in a Baer subline;*
2. *every Baer subline meets $U$ in $0, 1, 2$, or $q + 1$ points.*

*Proof.* This follows immediately from the fact that any three collinear points in $\mathrm{PG}(2, q^2)$ lie in a unique Baer subline. Hence, if $b$ is a set of points on a line $\ell$ in $\mathrm{PG}(2, q^2)$, then $b$ is a Baer subline of $\ell$ if and only if every Baer subline of $\ell$ meets $b$ in $0, 1, 2$, or $q + 1$ points.  □

Now let $U$ be a unital in $\mathrm{PG}(2, q^2)$, $q > 2$, such that every secant of $U$ meets $U$ in a Baer subline. Then by Lemma 7.19, every Baer subline meets $U$ in $0, 1, 2$, or $q + 1$ points. Thus Theorem 7.1 implies that $\mathcal{U}$ is ovoidal-Buekenhout-Metz with respect to every tangent line. Note that Theorem 7.3 now could be used to immediately deduce that $U$ is classical. However, we prefer to present the original proof given by Lefèvre-Percsy which does not rely on the powerful characterization of elliptic quadrics given in Theorem 7.18.

We begin by proving a weaker version of the desired theorem.

**Lemma 7.20.** *Let $U$ be a unital of $\mathrm{PG}(2, q^2)$, $q > 2$, such that every secant of $U$ is a Baer subline. Then $U$ is an orthogonal-Buekenhout-Metz unital.*

*Proof.* By Theorem 7.1, $\mathcal{U}$ is an ovoidal-Buekenhout-Metz unital with respect to every tangent line. Thus we let $\ell_\infty$ be the tangent line to $U$ at some point $P_\infty$. Then in the Bruck-Bose representation, $U$ is represented in $\mathrm{PG}(4, q)$ by an ovoidal cone $\mathcal{U}$ with vertex $V$ on the spread line $p_\infty$ and with base ovoid $\mathcal{O}$ in some hyperplane $\Sigma$. Note that $\Sigma$ meets $p_\infty$ in some point $Q \neq V$, and thus $\Sigma$ contains a unique spread line $m_\Sigma \neq p_\infty$ (see Figure 7.3).

Our aim is to force $\mathcal{O}$ to be an elliptic quadric by showing a sufficient number of oval plane sections of $\mathcal{O}$ are conics. (Again, we will not use the

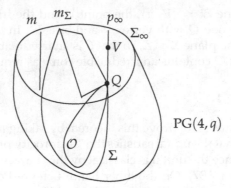

**Fig. 7.3.** Proof of Lemma 7.20

powerful result in [68] that states only one conic section is needed to force the ovoid to be an elliptic quadric.)

First we consider the planes in $\Sigma$ which pass through the spread line $m_\Sigma$. There are $q - 1$ such planes which meet $\mathcal{O}$ in an oval section. These planes correspond to secant lines of $U$ in $PG(2, q^2)$, which by assumption meet $U$ in various Baer sublines. Thus from the Bruck-Bose correspondence we know that each of these $q - 1$ planes meets $\mathcal{O}$ in a conic, thereby yielding $q - 1$ conic sections of $\mathcal{O}$.

We now consider other planes of $\Sigma$ that meet $\mathcal{O}$ in an oval. Let $m$ be a line of the spread $\mathcal{S}$ distinct from both $p_\infty$ and $m_\Sigma$. There are $q + 1$ hyperplanes containing the plane determined by $m$ and $V$, and each of these hyperplanes meets $\Sigma$ in a plane through the line $\langle V, m \rangle \cap \Sigma$. We investigate how these hyperplanes meet $\mathcal{O}$. Of course, one of these hyperplanes is the hyperplane at infinity, and it meets $\mathcal{O}$ only in the point $Q$.

The spread line $m$ corresponds to a point $M$ on $\ell_\infty$ in $PG(2, q^2)$. If $\ell$ is a line of $PG(2, q^2)$ through $M$ that is secant to $U$, then $\ell$ meets $U$ in a Baer subline by assumption. As this Baer subline has no point on $\ell_\infty$, we know that $\ell$ corresponds to a plane $\pi$ through $m$ that meets $\mathcal{U}$ in a conic. If $\Gamma = \langle \pi, V \rangle$, then $\Gamma$ is a hyperplane meeting $\mathcal{U}$ in a quadratic cone with vertex $V$ and base the conic $\pi \cap \mathcal{U}$. This cone necessarily meets $\Sigma$ in a conic of the plane $\Sigma \cap \Gamma$. Allowing $\ell$ to vary over the $q^2 - q$ secant lines to $U$ through $M$, we obtain $q - 1$ distinct conic sections of $\mathcal{O}$ in this way. Namely, each plane of $\Sigma$ through the line $\langle V, m \rangle \cap \Sigma$, except the two tangent planes to $\mathcal{O}$ through this exterior line, intersects $\mathcal{O}$ in a conic. Thus we also generate $q - 1$ conics of $\mathcal{O}$ for each spread line $m$ distinct from $p_\infty$ and $m_\Sigma$.

Now, any two distinct lines $m, m'$ of $\mathcal{S} \setminus \{p_\infty, m_\Sigma\}$ meet $\Sigma$ in two distinct points. Furthermore, the two lines $\langle V, m \rangle \cap \Sigma$ and $\langle V, m' \rangle \cap \Sigma$ are distinct. Hence, as $m$ ranges over the $q^2 - 1$ lines of $\mathcal{S} \setminus \{p_\infty, m_\Sigma\}$, the line $\langle V, m \rangle \cap \Sigma$ ranges over the $q^2 - 1$ lines of $\Sigma_\infty \cap \Sigma$ distinct from $m_\Sigma$ and not on the point $Q = p_\infty \cap \Sigma$. Each of these $q^2 - 1$ lines yields $q - 1$ conics of $\mathcal{O}$, and these conics are all distinct. Thus, we obtain $(q^2 - 1)(q - 1) = q^3 - q^2 - q + 1$ conics on $\mathcal{O}$ in this way.

Adding in the conics we obtained from the planes through $m_\Sigma$, we have shown the existence of

$$(q^3 - q^2 - q + 1) + (q - 1) = q^3 - q^2$$

distinct conics on the ovoid $\mathcal{O}$. Using a result of Barlotti [29] or a result of Segre [196], this is sufficient information to deduce that $\mathcal{O}$ is an elliptic quadric and hence $U$ is an orthogonal-Buekenhout-Metz unital, proving the lemma.    □

The characterization of orthogonal-Buekenhout-Metz unitals in Theorem 7.2 now implies that $U$ is classical as required.

It should be noted that the result in [196] shows that if an ovoid $\mathcal{O}$ has at least $(q^3 - q^2 + 2q)/2$ plane sections that are conics, then $\mathcal{O}$ is an elliptic quadric. The weaker result in [29], which states that if at least $q^3 - q^2$ plane sections of an ovoid $\mathcal{O}$ are conics, then $\mathcal{O}$ is an elliptic quadric, is the result that was used by Lefèvre-Percsy in her original proof. However, this result appears in some lecture notes of Barlotti, and may not be easily accessible. Nonetheless, the stronger (and earlier) result of Segre is all that is needed.

**Proof of Theorem 7.5**

Let $U$ be a unital in $\mathrm{PG}(2, q^2)$ such that $\ell_\infty$ is secant to $U$. Suppose that every Baer subline with a point on $\ell_\infty$ meets $U$ in 0, 1, 2, or $q + 1$ points. We show that $U$ is classical by using the Bruck-Bose representation, following the proof given in [33]. In particular, we show that $U$ is a nonsingular-Buekenhout unital by showing that $U$ corresponds to a nonsingular quadric in the Bruck-Bose representation. We then use the characterization of Theorem 4.5 to deduce that $U$ is a classical unital.

In the Bruck-Bose representation, $U$ corresponds to a set of points $\mathcal{U}$ of $\mathrm{PG}(4, q)$ that meets $\Sigma_\infty$ in a hyperbolic quadric, one of whose ruling families of lines consists of $q + 1$ lines of the spread $\mathcal{S}$. We note that by the hypothesis, $\ell_\infty$ meets $U$ in a Baer subline, so that the above set of $q + 1$ lines in $\mathcal{S}$ forms a regulus. We will need the following preliminary lemmas.

**Lemma 7.21.** *Using the above notation, $\mathcal{U}$ is a set of type $(0, 1, 2, q + 1)$.*

*Proof.* Let $P, Q$ be two points of $\mathcal{U}$. We show that either $PQ \cap \mathcal{U} = \{P, Q\}$ or the line $PQ$ is contained in $\mathcal{U}$. If both $P$ and $Q$ lie in $\Sigma_\infty$, then the line $PQ$ meets the hyperbolic quadric $\mathcal{U} \cap \Sigma_\infty$ in 0, 1, 2, or $q + 1$ points, and thus $PQ$ meets $\mathcal{U}$ in two or all of its $q + 1$ points.

We now suppose that at least one of the points $P$ and $Q$ is not in $\Sigma_\infty$. Let $R$ be the point $PQ \cap \Sigma_\infty$ ($R$ may be equal to $P$ or $Q$). Let $m_R$ be the line of $\mathcal{S}$ passing through $R$ and let $\pi$ be the plane containing $PQ$ and $m_R$. Then $\pi$

corresponds to a line $\ell$ in $\mathrm{PG}(2, q^2)$ and the line $PQ$ corresponds to a Baer subline of $\ell$ with a point on $\ell_\infty$. By the hypothesis of the theorem, this Baer subline intersects $U$ in $0, 1, 2$, or $q + 1$ points. Hence in $\mathrm{PG}(4, q)$ the line $PQ$ intersects $U$ in two or all of its points. $\qquad\square$

**Lemma 7.22.** $U$ *contains no plane of* $\mathrm{PG}(4, q)$.

*Proof.* Denote the lines of $S$ in $U \cap \Sigma_\infty$ by $t_0, t_1, \ldots, t_q$. So in $\mathrm{PG}(2, q^2)$, $U$ meets $\ell_\infty$ in the points $T_0, T_1, \ldots, T_q$, where the line $t_i \in S$ corresponds to the point $T_i \in \ell_\infty$ for $i = 0, 1, \ldots, q$.

We first show that $U$ contains no plane through one of the $t_i$. Let $\pi$ be a plane of $\mathrm{PG}(4, q)$ that meets $\Sigma_\infty$ in the line $t_i$. We will show that $\pi$ meets $U$ either in $t_i$ or in $t_i$ and another line $m$. The plane $\pi$ corresponds to a line $\ell$ of $\mathrm{PG}(2, q^2)$ that contains $T_i$ and is either tangent or secant to $U$. If $\ell$ is tangent to $U$, then $\ell \cap U = T_i$ and thus $\pi \cap U = t_i$. If $\ell$ is secant to $U$, then $\ell \cap U$ consists of $q + 1$ points, one of which is $T_i$. By the hypothesis of the theorem these points form a Baer subline of $\mathrm{PG}(2, q^2)$. This Baer subline corresponds to a line $m$ of $\mathrm{PG}(4, q)$ that meets $\Sigma_\infty$ in a point of $t_i$. Thus $\pi$ meets $U$ in the lines $m$ and $t_i$. Hence $U$ contains no plane through the lines $t_0, t_1, \ldots, t_q$ of the spread $S$.

Now suppose $U$ contains a plane $\beta$ not through one of the $t_i$. This plane intersects $\Sigma_\infty$ in a line $m$ contained in $U$, $m \neq t_i$ for $i = 0, 1, \ldots, q$, and hence $m$ meets each $t_i$. Thus $\beta$ corresponds to a Baer subplane of $\mathrm{PG}(2, q^2)$ that is contained in $U$. This contradicts Theorem 6.1 which says that the maximum number of points a unital and a Baer subplane can have in common is $2(q + 1)$. Hence $U$ contains no plane of $\mathrm{PG}(4, q)$. $\qquad\square$

**Lemma 7.23.** *Let* $U$ *be a unital in* $\mathrm{PG}(2, q^2)$, $q > 2$, *secant to* $\ell_\infty$. *If all Baer sublines having a point on* $\ell_\infty$ *intersect* $U$ *in* $0, 1, 2$, *or* $q + 1$ *points, then* $U$ *is a nonsingular-Buekenhout unital with respect to* $\ell_\infty$.

*Proof.* By Lemmas 7.21 and 7.22, $U$ is a set of type $(0, 1, 2, q + 1)$ that contains no plane. Further, $U$ contains $q^3 - q$ affine points and $q + 1$ points on $\ell_\infty$, so that $|U| = (q^3 - q) + (q + 1)^2 = q^3 + q^2 + q + 1$. Hence we can apply the characterization in Theorem 7.13. That is, $U$ is either a nonsingular quadric $\mathcal{P}_4$, or $\mathcal{P}_4$ union its nucleus. As $\mathcal{P}_4$ has $q^3 + q^2 + q + 1$ points, $U$ is a nonsingular quadric $\mathcal{P}_4$ in $\mathrm{PG}(4, q)$. Furthermore, $U$ meets $\Sigma_\infty$ in a hyperbolic quadric that contains a regulus of $S$. Hence $U$ is a nonsingular-Buekenhout unital of $\mathrm{PG}(2, q^2)$ with respect to $\ell_\infty$. $\qquad\square$

Lemma 7.23 and Theorem 4.5 immediately give the characterization we require.

Finally, we note that the same proof generalizes Lemma 7.23 to give a characterization of nonsingular-Buekenhout unitals in translation planes of order $q^2$ with kernel containing $\mathrm{GF}(q)$.

**Proof of Theorem 7.6**

Let $U$ be a unital in $\mathrm{PG}(2, q^2)$ tangent to $\ell_\infty$ in the point $T$. Suppose that the $q^2$ secants to $U$ through $T$ meet $U$ in a Baer subline. In $\mathrm{PG}(4, q)$ this means $\mathcal{U}$ is the union of the spread line $t$ and $q^2$ lines $\ell_1, \ell_2, \ldots, \ell_{q^2}$ of $\mathrm{PG}(4, q) \backslash \Sigma_\infty$ each meeting $t$ in a point, but pairwise having no common point in $\mathrm{PG}(4, q) \backslash \Sigma_\infty$. We prove the result using a series of lemmas, following the proof given in [87] and [183].

**Lemma 7.24.** *Let $\pi$ be a plane of $\mathrm{PG}(4, q) \backslash \Sigma_\infty$ passing through $t$. Then $\pi$ meets $\mathcal{U}$ in $t$ and one of the lines $\ell_i$.*

*Proof.* The plane $\pi$ corresponds to a line $\ell$ of $\mathrm{PG}(2, q^2)$ meeting $\ell_\infty$ in $T$. By the hypothesis, $\ell$ meets $U$ in $q + 1$ points that form a Baer subline. Hence $\pi \cap \mathcal{U}$ is the union of $t$ and one of the lines $\ell_i$ of $\mathrm{PG}(4, q) \backslash \Sigma_\infty$. $\quad\square$

**Lemma 7.25.** *If two distinct lines $\ell_i$ and $\ell_j$ meet, then the plane they span contains no further point of $\mathcal{U}$.*

*Proof.* If $\ell_i$ and $\ell_j$ meet, then they meet in a point of $t$. The plane $\beta = \langle \ell_i, \ell_j \rangle$ does not contain $t$, otherwise it would correspond to a line of $\mathrm{PG}(2, q^2)$ containing more than $q + 1$ points of $U$. Hence $\beta$ meets $\Sigma_\infty$ in a line not in $S$ and thus corresponds to a Baer subplane $\mathcal{B}$ of $\mathrm{PG}(2, q^2)$. Since $\mathcal{B}$ contains at least one tangent of $U$, namely, $\ell_\infty$, by Theorem 6.1 we have that $|\mathcal{B} \cap U| \leq 2q + 1$. Hence $\mathcal{B}$ contains at most $2q$ affine points of $U$, and thus $\beta$ contains no further point of $\mathcal{U}$. $\quad\square$

**Lemma 7.26.** *Let $\ell_i, \ell_j$ be distinct lines that meet in a point $V$ of $t$. Let $\beta = \langle \ell_i, \ell_j \rangle$ and $\Sigma = \langle \beta, t \rangle$. Then $\Sigma$ meets $\mathcal{U}$ in a cone projecting an oval from the point $V$.*

*Proof.* We first show that $q$ of the lines in $\{\ell_1, \ell_2, \ldots, \ell_{q^2}\}$ pass through $V$. By Lemma 7.24, the $q$ planes of $\mathrm{PG}(4, q) \backslash \Sigma_\infty$ in $\Sigma$ which pass through $t$ each contain a distinct line $\ell_k$. If one of these lines did not pass through $V$, then it meets $\beta$ in a point not on $\ell_i$ or $\ell_j$, contradicting Lemma 7.25. Note that if $\Sigma$ contains a point of $\ell_r \setminus t$, then $\Sigma$ contains the line $\ell_r$. Hence $\Sigma$ meets $\mathcal{U}$ in the line $t$ and $q$ of the lines in $\{\ell_1, \ell_2, \ldots, \ell_{q^2}\}$. To show that this cone is an oval cone, we need to show that every plane of $\Sigma$ through $V$ contains at most two of these lines. By Lemma 7.24, a plane through $t$ contains at most one line from $\ell_1, \ell_2, \ldots, \ell_{q^2}$, and by Lemma 7.25, a plane not through $t$ contains at most two lines from $\ell_1, \ell_2, \ldots, \ell_{q^2}$. The result now follows. $\quad\square$

**Lemma 7.27.** *Let $\ell_i, \ell_j$ be distinct lines that meet in a point $V$ of $t$. Let $\beta = \langle \ell_i, \ell_j \rangle$ and $\Sigma = \langle \beta, t \rangle$. By Lemma 7.26, $\Sigma \cap \mathcal{U}$ is a cone with vertex $V$. If a further line $\ell_k$ of $\mathcal{U}$ passes through $V$, then all of $\ell_1, \ell_2, \ldots, \ell_{q^2}$ pass through $V$.*

*Proof.* Without loss of generality, suppose that $t, \ell_1, \ell_2, \ldots, \ell_q$ are the $q + 1$ lines of the cone $\Sigma \cap \mathcal{U}$, and that $\ell_{q+1}$ also passes through $V$. For each $i = 1, 2, \ldots, q$, the space $\langle t, \ell_{q+1}, \ell_i \rangle$ is a three-dimensional space, which meets $\mathcal{U}$ in a cone with vertex $V$ by Lemma 7.26. Hence $q(q-1) + 1 = q^2 - q + 1$ of the lines $\ell_1, \ell_2, \ldots, \ell_{q^2}$ pass through $V$, leaving $q - 1$ further lines from $\{\ell_1, \ell_2, \ldots, \ell_{q^2}\}$ lying in the remaining 3-space $\Sigma'$ through the plane $\langle t, \ell_{q+1} \rangle$.

To complete the proof, we work in $\mathcal{Q}$, the quotient space of $PG(4, q)$ with respect to $t$. The *points* of $\mathcal{Q}$ are the planes through $t$, and the *lines* of $\mathcal{Q}$ are the 3-spaces through $t$. Incidence is containment, and $\mathcal{Q}$ is a projective plane of order $q$. A point of $\mathcal{Q}$ is called *full* if it corresponds to a plane of $PG(4, q) \backslash \Sigma_\infty$ through $t$ that contains a line of $\mathcal{U}$ through $V$. The 3-space $\Sigma_\infty$ corresponds to a line of $\mathcal{Q}$, none of whose points are full. The space $\Sigma = \langle \beta, t \rangle$ corresponds to a line of $\mathcal{Q}$, all of whose points apart from $P = \Sigma_\infty \cap \Sigma$ are full. The plane $\alpha = \langle t, \ell_{q+1} \rangle$ corresponds to a full point $\alpha$ of $\mathcal{Q}$. The 3-space $\Sigma'$ defined in the previous paragraph corresponds to a line $\Sigma'$ of $\mathcal{Q}$ with one full point $\alpha$.

Thus all points of $\mathcal{Q}$ are full except the points of $\Sigma_\infty$ and possibly $\Sigma' \backslash \alpha$. By Lemma 7.26, if two points of $\mathcal{Q} \backslash \Sigma_\infty$ are full, then every point of $\mathcal{Q} \backslash \Sigma_\infty$ on the line joining them is full. Thus the set of full points of $\mathcal{Q}$ is an affine subspace of $\mathcal{Q}$, and hence is either empty, a single point, the points of $\mathcal{Q} \backslash \Sigma_\infty$ on a line of $\mathcal{Q}$ distinct from $\Sigma_\infty$, or all of $\mathcal{Q} \backslash \Sigma_\infty$. Therefore the set of full points of $\mathcal{Q}$ must be all of $\mathcal{Q} \backslash \Sigma_\infty$. Thus every plane of $PG(4, q) \backslash \Sigma_\infty$ meets $\mathcal{U}$ in a line $\ell_k$ through $V$ for some $k$. That is, we have shown that the $q^2$ lines $\ell_1, \ell_2, \ldots, \ell_{q^2}$ all pass through $V$. □

**Lemma 7.28.** *If not all of the lines $\ell_1, \ell_2, \ldots, \ell_{q^2}$ pass through the same point of $t$, then they fall into $q$ cones $C_1, C_2, \ldots, C_q$ in 3-spaces $\Sigma_1, \Sigma_2, \ldots, \Sigma_q$ with distinct vertices $V_1, V_2, \ldots, V_q$ on $t$. Furthermore, the common intersection of the spaces $\Sigma_1, \Sigma_2, \ldots, \Sigma_q$ is a plane $\alpha$ tangent to each cone $C_i$ in $\Sigma_i$.*

*Proof.* By Lemma 7.26, if two of the $\ell_i$ pass through a point of $t$, then $q$ of them do. By Lemma 7.27, if the $\ell_i$ do not all pass through the same point of $t$, then no more than $q$ lines pass through each point of $t$. Hence the lines $\ell_1, \ell_2, \ldots, \ell_{q^2}$ form $q$ cones $C_1, C_2, \ldots, C_q$ with distinct vertices on $t$. Now consider the plane $\alpha = \Sigma_\infty \cap \Sigma_i$ for some $i$. Since $\mathcal{U} \cap \Sigma_\infty = t$, it follows that $\alpha$ is tangent to each cone $C_1, C_2, \ldots, C_q$ and $\alpha \subseteq \Sigma_i$ for all $i$. □

We now show that the situation in Lemma 7.28 cannot occur. We give the proof for $q > 3$ given in [183]. We refer the reader to [87, Lemma 3.8] for the proof in the case $q = 3$.

**Lemma 7.29.** *If $q > 3$, all the lines $\ell_1, \ell_2, \ldots, \ell_{q^2}$ pass through a common point $V$ on $t$.*

*Proof.* Suppose not, thus $\mathcal{U}$ has the cone structure described in Lemma 7.28. Let $\pi$ be a plane of $PG(4, q) \backslash \Sigma_\infty$ through some spread line $m$ that meets two lines, say $\ell_1$ and $\ell_2$, from distinct cones of $\mathcal{U}$. Thus $\pi$ corresponds to a secant line $\ell$ of $U$ that meets $\ell_\infty$ in the point $M$. Let $P_1 = \ell_1 \cap \pi$ and $P_2 = \ell_2 \cap \pi$. In $\mathrm{PG}(2, q^2)$, there are $q$ Baer sublines of $\ell$ that contain $P_1$ and $P_2$, but not $M$ (this follows from Theorem 2.6 as three distinct points lie on a unique Baer subline). Hence there are $q$ Baer conics in $\pi \backslash m$ containing $P_1$ and $P_2$. By Theorem 3.21, for each of these Baer conics there are $q + 1$ Baer ruled cubics containing the Baer conic as well as the line $t$. Each such Baer ruled cubic is determined completely by joining $P_1$ with one of the points on $t$. Hence there are $q$ Baer ruled cubics containing $\ell_1$ and $P_2$. As a quadrangle is contained in a unique Baer subplane (Theorem 2.8), no two Baer ruled cubics through $\ell_1$ and $P_2$ contain the same generator line through $P_2$. Hence there exists a unique Baer ruled cubic $\mathcal{B}$ containing the lines $\ell_1$ and $\ell_2$ of $U$. Let $C$ be the Baer conic $\pi \cap \mathcal{B}$, and let $\mathcal{B}$ be the Baer subplane of $\mathrm{PG}(2, q^2)$ represented by $\mathcal{B}$. Label the generators of $\mathcal{B}$ by $\ell_1, \ell_2, m_3, m_4, \ldots, m_{q+1}$ such that they meet $t$ in the points $V_1, V_2, \ldots, V_q, Q$, respectively, where $V_i$ are the vertices of the $q$ cones and $Q$ is the unique nonvertex point on $t$. Note that $m_3, m_4, \ldots, m_{q+1}$ are not lines of $U$ since by Theorem 6.1, $\mathcal{B}$ can intersect $U$ in at most $2q + 2$ points.

Consider the plane $\langle m_{q+1}, t \rangle$. By Lemma 7.24, it contains exactly one line $\ell_i$ of $U$ ($i \neq 1, 2$). Note that $\ell_i \cap m_{q+1}$ is an affine point of $U$. Thus we know that $\mathcal{B} \cap U$ contains $\ell_1, \ell_2, t$ and $\ell_i \cap m_{q+1}$. This corresponds to $q + q + 1 + 1 = 2q + 2$ points of $\mathcal{B} \cap U$. By Theorem 6.1, there is no further point in $\mathcal{B} \cap U$. Hence no line $m_3, m_4, \ldots, m_q$ can intersect $U$ in an affine point.

The plane $\langle m_i, t \rangle$ ($i = 3, 4, \ldots, q + 1$) contains a line $\ell_j$ of $U$ ($j \neq 1, 2$). As $m_i$ passes through the vertex $V_i$, $\ell_j$ also passes through the vertex $V_i$. Recall that $V_i$ is the vertex of the cone $\mathcal{C}_i$ which lies in the 3-space $\Sigma_i$. Hence the $q$ generators $\ell_1, \ell_2, m_3, m_4, \ldots, m_q$ of $\mathcal{B}$ lie in the cone spaces $\Sigma_1, \Sigma_2, \ldots, \Sigma_q$, respectively. Now $\alpha$ is the plane of common intersection of $\Sigma_1, \Sigma_2, \ldots, \Sigma_q$. Let $P = \pi \cap \alpha$. As $\alpha \subset \Sigma_\infty$, we have $P = m \cap \alpha$. But $m$ is external to the Baer conic $C = \pi \cap \mathcal{B}$, and hence $P$ is not the nucleus of the conic $C$ if $q$ is even. As $q > 3$, there exists a chord of the conic $C$ through the point $P$ incident with $C$ in two distinct points $Q_1$ and $Q_2$, neither of which is on $m_{q+1}$. So $Q_1$ and $Q_2$ lie on two distinct generators of $\mathcal{B}$, and hence belong to two distinct cone spaces, say $\Sigma_i$ and $\Sigma_j$. Thus both $Q_1$ and $Q_2$ belong to $\alpha$, a contradiction as $\alpha \subset \Sigma_\infty$. Thus all of the lines $\ell_1, \ell_2, \ldots, \ell_{q^2}$ pass through a common point $V$ on $t$. □

Note that if $q = 3$, there may be no chord of $C$ through $P$ that does not intersect $m_{q+1}$, and in that case the chord may lie in a unique cone space. Hence this proof is not valid in the case $q = 3$.

Now Lemmas 7.27 and 7.28 imply that $U$ is the union of $q$ cones with either a common vertex on $t$ or $q$ distinct vertices on $t$. Furthermore, the second case cannot occur by Lemma 7.29, and hence $U$ is the union of $q$ cones with a common vertex on $t$. The next lemma shows that such a set is an ovoidal cone. This will complete the proof of Theorem 7.6.

**Lemma 7.30.** *If all the lines* $\ell_1, \ell_2, \ldots, \ell_{q^2}$ *pass through a common point* $V$ *of* $t$, *then* $U$ *is an ovoidal-Buekenhout-Metz unital.*

*Proof.* We have $\mathcal{U}$ is a set of $q^2 + 1$ lines. Let $\Sigma$ be a 3-space that meets $t$ in a point distinct from $V$. Then the $q^2 + 1$ lines $\ell_1, \ell_2, \ldots, \ell_{q^2}, t$ meet $\Sigma$ in a set $\mathcal{O}$ containing $q^2 + 1$ points. By Lemmas 7.24 and 7.25, we know that no plane through $V$ meets $\mathcal{U}$ in more than two lines. Hence no line in $\Sigma$ meets $\mathcal{O}$ in more than two points. As $q > 2$, this implies that $\mathcal{O}$ is an ovoid. Hence $\mathcal{U}$ is an ovoidal cone and therefore $U$ is an ovoidal-Buekenhout-Metz unital. $\quad\square$

## 7.3 Other Configurational Characterizations

### 7.3.1 Tallini Scafati Characterizations

Tallini Scafati [192] studied sets of points in finite projective planes with certain prescribed intersection numbers, and obtained several results related to unitals. In this section we present those results. Let $\mathcal{P}$ be a projective plane of order $q$, and let $m$ and $n$ be distinct positive integers with $1 \leq m < n \leq q+1$. We say that a point set $\mathcal{K}$ in $\mathcal{P}$ is of **type** $(m, n)$ if every line of $\mathcal{P}$ meets $\mathcal{K}$ in $m$ or $n$ points and, moreover, both intersection sizes occur. As usual, a line meeting $\mathcal{K}$ in $m$ points will be called an *m-secant*. Also, we say that $\mathcal{K}$ is a **k-set** if $\mathcal{K}$ has $k$ points. One of the main results of Tallini Scafati on these sets is showing that a set of type $(1, n)$ in any projective plane of prime power order is either a unital or a Baer subplane. To prove this result, we begin with some counting.

**Lemma 7.31.** *Let* $\mathcal{K}$ *be a k-set of type* $(m, n)$ *in a projective plane* $\mathcal{P}$ *of order* $q$. *Let* $t_m$ *be the total number of m-secants of* $\mathcal{K}$ *and let* $t_n$ *be the total number of n-secants of* $\mathcal{K}$ *in* $\mathcal{P}$. *Then there are four constants associated with* $\mathcal{K}$, *these are denoted as follows:* $u_m$ *is the number of m-secants through a point* $Q \notin \mathcal{K}$, $u_n$ *is the number of n-secants through a point* $Q \notin \mathcal{K}$, $v_m$ *is the number of m-secants through a point* $P \in \mathcal{K}$, $v_n$ *is the number of n-secants through a point* $P \in \mathcal{K}$. *Moreover, the following equations hold:*

$$t_m = \frac{n(q^2 + q + 1) - k(q+1)}{n - m}, \quad t_n = \frac{k(q+1) - m(q^2 + q + 1)}{n - m}, \quad (7.1)$$

$$u_m = \frac{n(q+1) - k}{n - m}, \quad u_n = \frac{k - m(q+1)}{n - m}, \quad (7.2)$$

$$v_m = \frac{n(q+1) - k - q}{n - m}, \quad v_n = \frac{k + q - m(q+1)}{n - m}, \quad (7.3)$$

$$m t_m = k v_m, \quad n t_n = k v_n. \quad (7.4)$$

*Proof.* Every line in the plane is either an $m$-secant or an $n$-secant of $\mathcal{K}$, hence $t_m + t_n = q^2 + q + 1$. Counting in two ways the incident pairs $\{X, \ell\}$, where $X$ is a point on the line $\ell$ and $X \in \mathcal{K}$, yields $mt_m + nt_n = k(q+1)$. Solving these two linear equations gives the required values for $t_m, t_n$ found in (7.1).

Now let $Q$ be any point not in $\mathcal{K}$. The $q + 1$ lines through $Q$, which are necessarily $m$-secants or $n$-secants, partition the $k$ points of $\mathcal{K}$. Solving simultaneously the resulting two linear equations shows that indeed there are a constant number of $m$-secants and $n$-secants through a point not in $\mathcal{K}$, independent of the particular choice of $Q$, and these constants are given in (7.2). Similarly, if $P$ is any point of $\mathcal{K}$, the $q + 1$ punctured lines through $P$ partition the $k - 1$ points of $\mathcal{K} \setminus P$, showing that there are a constant number of $m$-secants and $n$-secants through any point of $\mathcal{K}$, independent of the choice of $P$, and these constants are given in (7.3).

Finally, counting in two ways the incident pairs $\{P, \ell\}$, where $P \in \mathcal{K}$ and $\ell$ is an $m$-secant yields the first equation found in (7.4). Repeating the count with $\ell$ an $n$-secant yields the second equation in (7.4).    □

We next prove a lemma bounding the number of points in $\mathcal{K}$.

**Lemma 7.32.** *Let $\mathcal{K}$ be a $k$-set of type $(m, n)$ in a projective plane $\mathcal{P}$ of order $q$, where we further assume $1 \le m < n \le q$ (that is, we no longer allow $n = q + 1$). Then*

$$n + mq \le k \le (n - 1)q + m.$$

*Proof.* Let $\ell$ be an $n$-secant of $\mathcal{K}$, and let $Q$ be a point on $\ell$ not in $\mathcal{K}$. This is possible as $n \le q$. Thus $u_n \ge 1$, and hence by (7.2) we have $k \ge n + mq$. For an upper bound, let $\ell$ be an $m$-secant of $\mathcal{K}$ and let $P \in \ell \cap \mathcal{K}$. Then the remaining $q$ lines through $P$ are either $m$-secants or $n$-secants. As $n > m$, the number of points of $\mathcal{K} \setminus \ell$ is at most $(n - 1)q$. That is, $k - m \le (n - 1)q$ and we have the desired upper bound on $\mathcal{K}$.    □

We now look at the structure of $\mathcal{K}$ in the two extremal cases of the above result.

**Lemma 7.33.** *Let $\mathcal{K}$ be a $k$-set of type $(m, n)$ in a projective plane $\mathcal{P}$ of prime power order $q = p^h$, where we further assume $1 \le m < n \le q$.*

1. *If $k = (n - 1)q + m$, then $\mathcal{K}$ is a unital or the complement of a Baer subplane.*
2. *If $k = n + mq$, then $\mathcal{K}$ is a Baer subplane or the complement of a unital.*

*Proof.* We prove (1) and note that a similar argument proves (2). Thus we suppose that $k = (n - 1)q + m$. Substituting this into the equations found in (7.3) yields $v_m = 1$ and $v_n = q$. Then substituting for $k$, $v_m$, and $v_n$ into the equations found in (7.4) yields

$$t_m = \frac{(n - 1)q}{m} + 1 \quad \text{and} \quad t_n = q^2 + \frac{q(m - q)}{n}.$$

Substituting into $t_m + t_n = q^2 + q + 1$ and simplifying then yields

$$n^2 - n(m+1) + m^2 - mq = 0. \tag{7.5}$$

Now from equations (7.2) and (7.3), we have $v_n - u_n = q/(n-m)$, and hence $n - m$ divides $q$. Let $q_1 = n - m$ and substitute into equation (7.5) to obtain

$$m^2 - m(q - q_1 + 1) + q_1(q_1 - 1) = 0.$$

Let the (integer) roots of this equation be $m_1$ and $m_2$, so that

$$m_1 m_2 = q_1(q_1 - 1), \tag{7.6}$$
$$m_1 + m_2 = q - q_1 + 1. \tag{7.7}$$

If $q_1 = 1$, then equations (7.6) and (7.7) imply that the two roots $m_1$ and $m_2$ are 0 and $q$. This implies that $m = q$, as we do not allow $m = 0$, and therefore $n = q + 1$, contradicting our assumption on $n$.

Hence we can assume $q_1 \neq 1$ and thus $q_1$ is a nonzero power of the prime $p$, where $q = p^h$. That is, $q_1 = p^i$ for some integer $i$, $1 \leq i \leq h$. Thus by equation (7.7), it is not possible for both $m_1$ and $m_2$ to be divisible by $p$. Therefore, without loss of generality, we may assume that $m_1$ is coprime to $p$ and hence, by equation (7.6), $m_2$ is a multiple of $q_1$. Let $m_2 = aq_1$, and thus by equation (7.6)

$$am_1 = q_1 - 1. \tag{7.8}$$

Substituting for $m_1, m_2$ into equation (7.7) and rewriting yields

$$q_1(a^2 - a(q/q_1 - 1) + 1) = a + 1.$$

Letting $b = a^2 - a(q/q_1 - 1) + 1$, we then have $q_1 b = a + 1$ and thus $a = q_1 b - 1$. Note that if $b = 0$, then $a = -1$ and so $m_2 = -q_1, m_1 = -q_1 + 1$. This further would imply by equation (7.7) that $q = -q_1$, a contradiction. Hence we know that $b \neq 0$. Substituting $a = q_1 b - 1$ into equation (7.8) yields

$$m_1 = \frac{q_1 - 1}{q_1 b - 1}.$$

Now $b, q_1$, and $m_1$ are integers and $b \neq 0$. Hence $b = 1$, implying that $m_1 = 1$ and $m_2 = q_1(q_1 - 1)$. Furthermore, substituting for $m_1$ and $m_2$ in equation (7.7) yields $1 + q_1^2 - q_1 = q - q_1 + 1$, and thus $q = q_1^2$.

Hence we have two possibilities for $m$. If $m = m_1 = 1$, then

$$n = q_1 + m = \sqrt{q} + 1 \quad \text{and} \quad k = (n-1)q + 1 = q\sqrt{q} + 1.$$

Hence $\mathcal{K}$ is a unital of $\mathcal{P}$ in this case. If $m = m_2 = q_1(q_1 - 1) = q - \sqrt{q}$, then

$$n = q \quad \text{and} \quad k = q^2 - \sqrt{q}.$$

Consider the complement $\overline{\mathcal{K}}$ of $\mathcal{K}$; namely, the points of $\overline{\mathcal{K}}$ are the points of $\mathcal{P}\backslash\mathcal{K}$. Then $\overline{\mathcal{K}}$ is a set of size $q^2 + q + 1 - (q^2 - \sqrt{q}) = q + \sqrt{q} + 1$, and every line of $\mathcal{P}$ meets $\overline{\mathcal{K}}$ in $q + 1 - m = \sqrt{q} + 1$ or $q + 1 - n = 1$ points. That is, $\overline{\mathcal{K}}$ is a $(q + \sqrt{q} + 1)$-set of type $(1, \sqrt{q} + 1)$. Hence the line joining any two points of $\overline{\mathcal{K}}$ is a $(\sqrt{q} + 1)$-secant of $\overline{\mathcal{K}}$. Furthermore, by equation (7.2), $u_{\sqrt{q}+1} = 1$ and thus two $(\sqrt{q} + 1)$-secants cannot meet in a point outside $\overline{\mathcal{K}}$. Therefore two such secants must meet in a point of $\overline{\mathcal{K}}$, and $\overline{\mathcal{K}}$ is a Baer subplane of $\mathcal{P}$. That is, $\mathcal{K}$ is the complement of a Baer subplane in the second case, proving the theorem. □

**Theorem 7.34.** *Let $\mathcal{K}$ be a $k$-set of type $(1, n)$ in a projective plane $\mathcal{P}$ of prime power order $q$, where we assume $1 < n \leq q$. Then $\mathcal{K}$ is either a unital or a Baer subplane of $\mathcal{P}$.*

*Proof.* We work in the dual plane $\mathcal{P}^*$ of $\mathcal{P}$. The $t_1$ tangents of $\mathcal{K}$ form a $t_1$-set $\mathcal{K}^*$ in $\mathcal{P}^*$. Each point of $\mathcal{P}$ lies on either $u_1$ or $v_1$ tangents of $\mathcal{K}$, and hence each line of $\mathcal{P}^*$ meets $\mathcal{K}^*$ in $u_1$ or $v_1$ points. That is, $\mathcal{K}^*$ is a $k'$-set of type $(m', n')$ in $\mathcal{P}^*$, where $k' = t_1, m' = v_1$, and $n' = u_1$. Hence, by equations (7.1), (7.2), and (7.3), we know that

$$k' = \frac{n(q^2 + q + 1) - k(q + 1)}{n - 1}, \quad m' = \frac{qn + n - k - q}{n - 1}, \quad n' = \frac{qn + n - k}{n - 1}.$$

Note that $m' = 0$ would imply that $k = nq + n - q$. Since $n > 1$, this would contradict the upper bound for $k$ given in Lemma 7.32. Similarly, $n' = q + 1$ would imply that $k = q + 1$, contradicting the lower bound for $k$ given in Lemma 7.32. Hence the parameters of $\mathcal{K}^*$ satisfy the conditions of Lemma 7.33. Since one may easily verify that $(n' - 1)q + m' = k'$, we know by Lemma 7.33 that either $\mathcal{K}^*$ is a unital (in which case $m' = 1, n' = \sqrt{q} + 1$) or $\mathcal{K}^*$ is the complement of a Baer subplane (in which case $m' = q - \sqrt{q}$, $n' = q$).

If $\mathcal{K}^*$ is a unital, then $m' = 1$ means $(qn + n - k - q)/(n - 1) = 1$ and rearranging yields

$$k = qn - q + 1.$$

Furthermore, $n' = \sqrt{q} + 1$ means $(qn + n - k)/(n - 1) = \sqrt{q} + 1$ and rearranging yields

$$k = qn - n\sqrt{q} + \sqrt{q} + 1.$$

Equating these two expressions for $k$ and rearranging yields $n = \sqrt{q} + 1$. Hence $k = q\sqrt{q} + 1$ and $\mathcal{K}$ is a unital of $\mathcal{P}$.

Similarly, if $\mathcal{K}^*$ is the complement of a Baer subplane, then $m' = q - \sqrt{q}$ and $n' = q$, implying $k = n + n\sqrt{q} - \sqrt{q}$ and $k = n + q$. Equating these two expressions for $k$ yields $n = \sqrt{q} + 1$. Hence $k = q + \sqrt{q} + 1$ and $\mathcal{K}$ is a Baer subplane of $\mathcal{P}$. □

Tallini Scafati [192] also investigated unitals satisfying the following reciprocity property.

**Definition 7.35.** *Let $U$ be a unital in a projective plane of order $q^2$. Let $P_1, P_2, P_3, P_4$ be a quadrangle of $U$ (that is, four points of $U$ with no three collinear), and let $\ell_1, \ell_2, \ell_3, \ell_4$ be the tangents to $U$ at these respective points. We say that $U$ satisfies the* **reciprocity property** *if whenever $\ell_1 \cap \ell_2 \in P_3 P_4$, then necessarily we have $\ell_3 \cap \ell_4 \in P_1 P_2$.*

Tallini Scafati proved that a unital in $PG(2, q^2)$ satisfying the reciprocity property is classical, a characterization which we now present. The first step is to show that a consequence of the reciprocity property is that the feet of any point $P \notin U$ are collinear. Recall that the *feet* of $P$ are the $q + 1$ points of $U$ lying on the $q + 1$ tangents to $U$ through $P$. In Corollary 2.4 we showed that if $U$ is classical, then the feet of any point $P \notin U$ are collinear.

**Lemma 7.36.** *Let $U$ be a unital in $PG(2, q^2)$ satisfying the reciprocity property. Then for each point $P \notin U$, the feet of $P$ are collinear.*

*Proof.* Suppose $q \geq 3$. Let $P$ be a point not in $U$, and let $\ell_1, \ell_2, \ell_3$ be three of the tangents through $P$. If $Q_1, Q_2, Q_3$ denote their respective points of tangency, it suffices to show that these points are collinear. Let $m$ be a secant through $P$, and suppose that $Q_1, Q_2, Q_3$ are not collinear. Then the lines $Q_1 Q_2$ and $Q_1 Q_3$ meet $m$ in distinct points (not necessarily on $U$). As $q + 1 \geq 4$, there are (at least) two points $R_1, R_2$ of $U$ on $m$ that are not incident with $Q_1 Q_2$ or $Q_1 Q_3$ (see Figure 7.4).

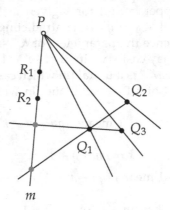

**Fig. 7.4.** Proof of Lemma 7.36

Thus we have two quadrangles of $U$, namely,

$$\{R_1, R_2, Q_1, Q_2\} \quad \text{and} \quad \{R_1, R_2, Q_1, Q_3\}.$$

Denote the tangents to $U$ through $R_1$ and $R_2$ by $m_1$ and $m_2$, respectively. Then the reciprocity property implies that $m_1 \cap m_2 \in Q_1 Q_2$ and $m_1 \cap m_2 \in Q_1 Q_3$. By assumption $Q_1, Q_2, Q_3$ are not collinear, and hence $m_1 \cap m_2 = Q_1$. This is a contradiction as $Q_1 \in U$ and $m_1$ is tangent to $U$ at $R_1 \neq Q_1$. Thus $Q_1, Q_2, Q_3$ must be collinear, and the result follows for $q \geq 3$. The case $q = 2$ is left as an exercise for the reader.                                                    □

**Theorem 7.37.** *Let $U$ be a unital in* $\mathrm{PG}(2, q^2)$. *If $U$ satisfies the reciprocity property, then $U$ is classical.*

*Proof.* Let $U$ be a unital in $\mathrm{PG}(2, q^2)$ satisfying the reciprocity property. We use Lemma 7.36 to define a polarity whose absolute points are the points of $U$. If $P \in U$, then define the *polar* of $P$ to be the tangent through $P$. If $P \notin U$, then define the *polar* of $P$ to be the line containing the $q + 1$ feet of $P$ (by Lemma 7.36 this line is well-defined). Let $\ell$ be a line of $\mathrm{PG}(2, q^2)$. If $\ell$ is a tangent to $U$, then define its *pole* to be the point of tangency. If $\ell$ is secant to $U$, then by Lemma 7.36 the tangents to the $q + 1$ points of $U$ on $\ell$ meet in a point $P$; we define $P$ to be the *pole* of $\ell$. Thus we have a well-defined bijection $\theta$ of order 2 between the points and lines of $\mathrm{PG}(2, q^2)$. To prove that this map is a polarity, we must prove that $\theta$ preserves incidence. That is, we must show that if a point $P$ lies on the line $\ell$, then the point $\ell^\theta$ lies on the line $P^\theta$.

Suppose first that $P$ is a point on the tangent line $\ell$, and let $\ell$ have point of tangency $\ell^\theta$. If $P \neq \ell^\theta$, then $P \notin U$. Thus $\ell^\theta$ is a foot of $P$ and the polar of $P$ is a line through $\ell^\theta$. Hence $\ell^\theta \in P^\theta$. Similarly, if $P = \ell^\theta$, then the polar of $P$ is $\ell$ and so $\ell^\theta \in P^\theta$.

Now suppose $P$ is a point on the secant line $\ell$. The pole of $\ell$ is the intersection of the tangents of the $q + 1$ points of $U$ on $\ell$. If $P \in U$, then the polar of $P$ is the tangent through $P$ and so contains the pole of $\ell$ as required. If $P \notin U$, then the polar of $P$ is a line $P^\theta$. As $\ell$ is not a tangent line and $P \notin U$, $P^\theta \cap \ell$ is a point not in $U$. Let $R_1$ and $R_2$ be two points of $P^\theta \cap U$, so that $\ell_1 = PR_1$ and $\ell_2 = PR_2$ are tangents of $U$. Let $Q_1$ and $Q_2$ be two points of $\ell \cap U$ with tangents $m_1$ and $m_2$, respectively. By construction, $m_1$ and $m_2$ contain the pole of $\ell$. Now $\{R_1, R_2, Q_1, Q_2\}$ is a quadrangle of $U$, and $\ell_1 \cap \ell_2 \in Q_1 Q_1$ (see Figure 7.5). Hence the reciprocity property implies that $m_1 \cap m_2 \in R_1 R_2$. That is, $\ell^\theta \in P^\theta$ as required.

Hence $\theta$ is a polarity. Note that the absolute points of $\theta$ are precisely the points of $U$, implying that $\theta$ is a unitary polarity of $\mathrm{PG}(2, q^2)$ and $U$ is a classical unital.                                                    □

### 7.3.2 Characterizations Using Feet

We now discuss some additional characterizations of the classical unital via feet. Recall from Corollary 2.4 that if $U$ is the classical unital, then the feet of all points $P \notin U$ are collinear. It was long conjectured that the converse of this must be true. In 1992, Thas [209] proved this conjecture, where he stated his result dually in the following equivalent way:

**Theorem 7.38.** *Let $U$ be a unital in* $\mathrm{PG}(2, q^2)$. *If the tangents of $U$ at collinear points of $U$ are concurrent, then $U$ is classical.*

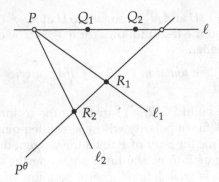

**Fig. 7.5.** Proof of Theorem 7.37

A main tool in Thas' proof is Segre's algebraic "Lemma of Tangents". That is, an algebraic argument is used to show that every secant of $U$ is a Baer subline, and hence one may deduce from Theorem 7.4 that the unital must be classical. An alternative proof of this characterization was given by Thas, Cameron, and Blokhuis [210]. The alternate approach uses some algebraic geometry, relying on the following characterization result due to Hirschfeld, Storme, Thas, and Voloch [132].

**Theorem 7.39.** *In* $\mathrm{PG}(2, q^2)$, $q \neq 2$, *any algebraic curve of degree* $q + 1$ *without linear components and with at least* $q^3 + 1$ *points in* $\mathrm{PG}(2, q^2)$ *must be a Hermitian curve.*

Characterizing the classical unital via feet was significantly improved by Aguglia and Ebert [10], who showed that if the feet of all points $P \notin U$ on *two* tangent lines of $U$ are collinear, then $U$ is classical. Their proof uses a similar algebraic technique to that used by Thas, beginning with the lemma stated below. The proof of Thas ([209, Lemma, page 98]) works under the milder hypothesis of this lemma, as we now illustrate.

**Lemma 7.40.** *Let* $U$ *be a unital in* $\mathrm{PG}(2, q^2)$, $q > 2$, *for which there are two points* $P_1$ *and* $P_2$ *in* $U$ *with tangent lines* $\ell_1$ *and* $\ell_2$, *respectively, such that for all points* $P \in \ell_1 \cup \ell_2$, $P \notin U$, *the feet of* $P$ *are collinear. Let* $P_3$ *be a point in* $U \setminus P_1 P_2$ *with tangent line* $\ell_3$. *Without loss of generality, choose coordinates such that* $E_1 = \ell_2 \cap \ell_3 = (1, 0, 0)$, $E_2 = \ell_3 \cap \ell_1 = (0, 1, 0)$, *and* $E_3 = \ell_1 \cap \ell_2 = (0, 0, 1)$. *If* $P_1 = (0, b, 1)$, $P_2 = (1, 0, c)$, *and* $P_3 = (a, 1, 0)$, *then* $(abc)^{q+1} = 1$ *(see Figure 7.6).*

*Proof.* By our hypotheses the $q + 1$ points of $P_2 P_3 \cap U$ are the feet of $E_1$. Other than $P_2$ and $P_3$, the remaining $q - 1$ feet of $E_1$ have coordinates which can be normalized so that the last coordinate is equal to 1. Let $y_1, y_2, \ldots, y_{q-1}$ be the middle coordinates of these $q - 1$ feet, so normalized. Note that these middle coordinates are all distinct and nonzero. Similarly, the $q - 1$ points

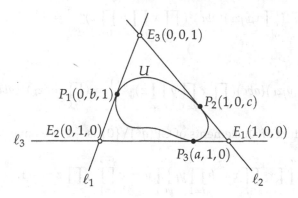

**Fig. 7.6.** The configuration of Lemma 7.40

of $P_1 P_3 \cap U$, other than $P_1$ and $P_3$, are the remaining feet of $E_2$ and can be normalized so that the first coordinate is 1. Let $z_1, z_2, \ldots, z_{q-1}$ be the third coordinates of these $q - 1$ feet, so normalized. Finally let $x_1, x_2, \ldots, x_{q-1}$ be the first coordinates of the $q - 1$ feet of $E_3$, other than $P_1$ and $P_2$ (and all on the line $P_1 P_2$), normalized so that the middle coordinate is 1. Let

$$X = \mathrm{GF}(q^2) \setminus \{0, a, x_1, x_2, \ldots, x_{q-1}\},$$
$$Y = \mathrm{GF}(q^2) \setminus \{0, b, y_1, y_2, \ldots, y_{q-1}\},$$
$$Z = \mathrm{GF}(q^2) \setminus \{0, c, z_1, z_2, \ldots, z_{q-1}\}.$$

Using dual coordinates, the tangent lines $\ell_1, \ell_2$, and $\ell_3$ to $U$ can be represented as $\ell_1 = (1, 0, 0)'$, $\ell_2 = (0, 1, 0)'$, and $\ell_3 = (0, 0, 1)'$. Hence any point $P$ of $U \setminus \{P_1, P_2, P_3\}$ has coordinates $P = (x, y, z)$ with each of $x, y, z$ nonzero. Thus the line $E_1 P$ has normalized dual coordinates of the form $(0, 1, -y/z)'$, where we think of $y/z$ as the "slope" $m_P^{(1)}$ of the line $E_1 P$. Similarly, we think of $E_2 P = (-z/x, 0, 1)'$ as having slope $m_P^{(2)} = z/x$ and $E_3 P = (1, -x/y, 0)'$ as having slope $m_P^{(3)} = x/y$. Thus $m_P^{(1)} m_P^{(2)} m_P^{(3)} = 1$ for all such points $P$ and hence

$$\prod_P m_P^{(1)} m_P^{(2)} m_P^{(3)} = 1. \tag{7.9}$$

Note that as $P$ varies over $U \setminus \{P_1, P_2, P_3\}$, the slope $m_P^{(1)}$ of $E_1 P$ takes on each value in $Y$ exactly $q + 1$ times, takes on each value $y_i$ exactly once, and takes on the value $b$ exactly $q$ times. Similarly, the slope $m_P^{(2)}$ of $E_2 P$ takes on each value in $Z$ exactly $q + 1$ times, takes on each value $z_i$ exactly once, and takes on the value $c$ exactly $q$ times. And the slope $m_P^{(3)}$ of $E_3 P$ takes on each value in $X$ exactly $q + 1$ times, takes on each value $x_i$ exactly once, and takes on the value $a$ exactly $q$ times. Hence from equation (7.9) we have

$$(\prod_{i,j,k} x_i y_j z_k)(abc)^q (\prod_{x\in X} x \prod_{y\in Y} y \prod_{z\in Z} z)^{q+1} = 1,$$

implying that

$$[(\prod_{i,j,k} x_i y_j z_k)(abc)(\prod_{x\in X} x \prod_{y\in Y} y \prod_{z\in Z} z)]^{q+1} = (\prod_{i,j,k} x_i y_j z_k)^q (abc).$$

As the product of all elements of $GF(q^2)\setminus\{0\}$ is $-1$, we have

$$a\prod_{i=1}^{q-1} x_i \prod_{x\in X} x = b\prod_{i=1}^{q-1} y_i \prod_{y\in Y} y = c\prod_{i=1}^{q-1} z_i \prod_{z\in Z} z = -1.$$

Combining the above equations, we obtain

$$(\prod_{i,j,k} x_i y_j z_k)^q (abc) = (-1)^{q+1} = 1$$

and therefore, raising both sides to the $q^{\text{th}}$ power, we have

$$(\prod_{i,j,k} x_i y_j z_k)(abc)^q = 1. \tag{7.10}$$

Choose new (normalized) coordinates in such a way that $P_1 = (1,0,0)$, $P_2 = (0,1,0)$, and $P_3 = (0,0,1)$. To accomplish this, one can simply apply the homography represented by the nonsingular matrix

$$A = \begin{pmatrix} -c & 1 & bc \\ ac & -a & 1 \\ 1 & ab & -b \end{pmatrix}$$

acting on points via right multiplication on the associated row vectors. As lines are represented by column vectors via dual coordinates, the action of this homography on lines is given via left multiplication by $A^{-1}$ on the associated column vectors.

One easily computes

$$A^{-1} = \frac{1}{1+abc} \begin{pmatrix} 0 & b & 1 \\ 1 & 0 & c \\ a & 1 & 0 \end{pmatrix},$$

observing that $abc + 1 \neq 0$ since $P_1, P_2, P_3$ are noncollinear. Using this inverse matrix, we see that the new (normalized) dual coordinates for the tangent lines $\ell_1, \ell_2$, and $\ell_3$ are $(0, a^{-1}, 1)'$, $(1, 0, b^{-1})'$, and $(c^{-1}, 1, 0)'$, respectively. Also the new normalized dual coordinates for tangent lines through $E_1, E_2, E_3$ are

$$(b - y_j, -cy_j, 1)', \quad (1, c - z_k, -az_k)', \quad (-bx_i, 1, x_i - a)',$$

respectively, as $i, j, k$ vary over $\{1, 2, \ldots, q-1\}$. Of course, as usual we are intentionally omitting the tangent lines $E_1 P_2$, $E_1 P_3$, $E_2 P_1$, $E_2 P_3$, $E_3 P_1$, and $E_3 P_2$.

We now repeat the above process to the dual unital of $U$ in the dual plane (still isomorphic to $PG(2, q^2)$), but using the new coordinate system. That is, $P_1, P_2, P_3$ are replaced by $\ell_1, \ell_2, \ell_3$, and the feet of $E_1, E_2, E_3$ are replaced by the tangent lines through $E_1, E_2, E_3$. Thus, using our computations above, equation (7.10) becomes

$$(a^{-1}b^{-1}c^{-1})^q \prod_i (-bx_i) \prod_j (-cy_j) \prod_k (-az_k) = 1.$$

Simplifying, we obtain

$$(abc)^{-1} \prod_{i,j,k} x_i y_j z_k = 1.$$

Comparing with equation (7.10), we have $(abc)^{q+1} = 1$ and the lemma is proved. $\qquad\square$

We now present the main result in [10].

**Theorem 7.41.** *Let $U$ be a unital in $PG(2, q^2)$, $q > 2$. Then $U$ is classical if and only if there are two points $P_1$ and $P_2$ in $U$ with tangent lines $\ell_1$ and $\ell_2$, respectively, such that for all points $P \in \ell_1 \cup \ell_2 \setminus \{P_1, P_2\}$, the feet of $P$ are collinear.*

*Proof.* Since classical unitals have the property that the feet of every exterior point are collinear, we only need to consider the converse. Let $U$ be a unital in $PG(2, q^2)$, $q > 2$, such that there are two points $P_1$ and $P_2$ in $U$ with tangent lines $\ell_1$ and $\ell_2$, respectively, such that for all points $P \in \ell_1 \cup \ell_2 \setminus \{P_1, P_2\}$, the feet of $P$ are collinear. We first show that $U$ is an ovoidal-Buekenhout-Metz unital.

Let $Q$ be a point in $\ell_2 \setminus \{P_2, \ell_1 \cap \ell_2\}$, let $\ell_3$ be a tangent line through $Q$ distinct from $\ell_2$, and let $P_3$ be its point of tangency. Note that $P_3 \notin P_1 P_2$ since the feet of $Q$ are the points of $P_2 P_3 \cap U$ and $P_1$ is not a foot of $Q$. We now choose normalized coordinates so that $Q = \ell_2 \cap \ell_3 = (1, 0, 0)$, $\ell_3 \cap \ell_1 = (0, 1, 0)$, and $\ell_1 \cap \ell_2 = (0, 0, 1)$ (see Figure 7.7). Then we may write $P_1 = (0, b, 1)$, $P_2 = (1, 0, c)$, and $P_3 = (a, 1, 0)$ for some nonzero elements $a, b, c \in GF(q^2)$. Hence by Lemma 7.40, we have

$$(abc)^{q+1} = 1. \tag{7.11}$$

The points of $P_2 P_3 \setminus \{P_2, P_3\}$ have coordinates of the form $P_2 + \lambda_i P_3$, for nonzero scalars $\lambda_i \in GF(q^2)$. In particular, we may assume the $q+1$ points of $P_2 P_3 \cap U$ have coordinates

$$F = \Big\{ R_i = (c^{-1} + \lambda_i a, \lambda_i, 1) \mid \lambda_i \in GF(q^2) \setminus \{0\}, i = 1, 2, \ldots, q \Big\}$$
$$\cup \{R_\infty = P_3 = (a, 1, 0)\},$$

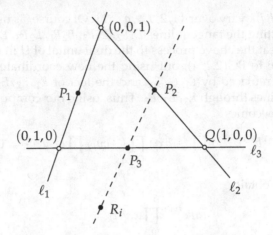

**Fig. 7.7.** The original coordinate system

where, without loss of generality, $\lambda_q = 0$ and so $R_q = P_2$. By choosing our reference quadrangle to be the four points $\ell_1 \cap \ell_2$, $\ell_1 \cap \ell_3$, $Q$, and $R_1$, we may further assume that the line joining $Q$ and $R_1$ has equation $y = z$, so that $R_1$ has coordinates $(c^{-1} + a, 1, 1)$. That is, we may assume without loss of generality that $\lambda_1 = 1$.

As $Q \in \ell_2$, our assumptions imply that the feet of $Q$ are collinear. Hence, as $F$ contains two of the feet of $Q$, necessarily $F$ consists of all the feet of $Q$. That is, the lines $QR_i$ all are tangents of $U$. If $R_i' = QR_i \cap \ell_1$ for $i = 1, 2, \ldots, q-1$, then $R_i' = (0, \lambda_i, 1)$ for $i = 1, 2, \ldots, q-1$.

Now fix some $i \in \{1, 2, \ldots, q-1\}$, and change the basis for coordinates so that $Q = (1, 0, 0)$, $\ell_1 \cap \ell_2 = (0, 0, 1)$, and $R_i' = (0, 1, 0)$ (see Figure 7.8).

Using the same conventions as in the proof of the previous lemma, the transformation matrix from old to new coordinates is

$$\begin{pmatrix} 1 & 0 & 0 \\ 0 & \lambda_i^{-1} & -\lambda_i^{-1} \\ 0 & 0 & 1 \end{pmatrix}.$$

In this new reference system, we have $P_1 = (0, b/(\lambda_i - b), 1)$, $P_2 = (1, 0, c)$, and $R_i = (c^{-1} + \lambda_i a, 1, 0)$. Note that since $P_1 \notin QR_i$, we have $\lambda_i - b \neq 0$. Applying Lemma 7.40 yields

$$\left( \frac{b + abc\lambda_i}{\lambda_i - b} \right)^{q+1} = 1.$$

Thus since $(x + y)^q = x^q + y^q$ for all $x, y \in GF(q^2)$, we have

$$ab^{q+1}c\lambda_i + a^q b^{q+1} c^q \lambda_i^q + (abc)^{q+1} \lambda_i^{q+1} = -b^q \lambda_i - b\lambda_i^q + \lambda_i^{q+1}. \qquad (7.12)$$

Using equation (7.11), we can simplify equation (7.12) to obtain

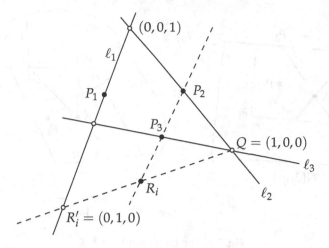

**Fig. 7.8.** The new coordinate system

$$\left(b^{q-1} + \lambda_i^{q-1}(abc + 1)^{q-1}\right)(abc + 1)b\lambda_i = 0.$$

Now $abc + 1 \neq 0$ as $P_1, P_2, P_3$ are not collinear, and hence we have

$$b^{q-1} + \lambda_i^{q-1}(abc + 1)^{q-1} = 0. \tag{7.13}$$

Putting $i = 1$, and recalling that $\lambda_1 = 1$, this equation becomes

$$b^{q-1} + (abc + 1)^{q-1} = 0.$$

Hence equation (7.13) becomes $b^{q-1}(1 - \lambda_i^{q-1}) = 0$. Thus $\lambda_i^{q-1} = 1$, and therefore $\lambda_i \in \mathrm{GF}(q)$ for $i = 1, 2, \ldots, q - 1$. That is, each $R_i$ is a $\mathrm{GF}(q)$-linear combination of $P_2$ and $P_3$.

Now let $P$ be a point in $\mathrm{PG}(2, q^2) \setminus P_2 P_3$ and consider the Baer subplane

$$B = \{dP_2 + eP_3 + fP \mid d, e, f \in \mathrm{GF}(q)\}.$$

By taking $d = 1$ and $f = 0$, it follows that the feet $F$ of $Q$ are contained in $B$. That is, $F$ is the intersection of the line $P_2 P_3$ and the Baer subplane $B$, so $F$ is a Baer subline.

As the point $Q$ varies over the points in $\ell_2 \setminus \{P_2, \ell_1 \cap \ell_2\}$, the line $P_2 P_3$ will vary over all the secants through $P_2$, except for $P_2 P_1$. Thus all the secants through $P_2$ (except possibly $P_2 P_1$) meet $U$ in Baer sublines. By interchanging $P_2$ and $P_1$, we similarly obtain that all the secants through $P_1$, except possibly $P_1 P_2$, also meet $U$ in Baer sublines. We now use a geometric argument to show that the secant $P_1 P_2$ meets $U$ in a Baer subline as well.

Let $\ell_2 = \ell_\infty$ be the line at infinity of $\mathrm{PG}(2, q^2)$. We consider the configuration shown in Figure 7.9. Let $m$ be a Baer subline contained in $U$ such

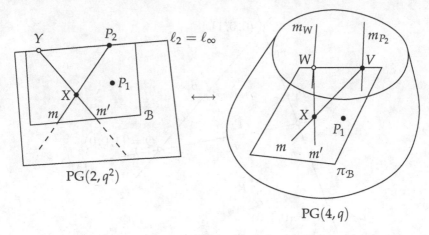

**Fig. 7.9.** The Baer subplane $\mathcal{B}$

that $P_2 \in m$ and $P_1 \notin m$. In the Bruck-Bose representation, the Baer subline $m$ corresponds to a line of $PG(4,q) \setminus \Sigma_\infty$ which we also denote by $m$. The line $m$ meets $\Sigma_\infty$ in a point $V$ of the spread line $m_{P_2}$ corresponding to the point $P_2$. Note the $P_1$ is a point of $AG(2,q^2)$ and so corresponds to a point $P_1$ of $PG(4,q) \setminus \Sigma_\infty$. Consider the plane $\pi_\mathcal{B} = \langle m, P_1 V \rangle$ of $PG(4,q)$. It does not contain the line $m_{P_2}$, since if it did, $\pi_\mathcal{B}$ would correspond to a line of $PG(2,q^2)$ containing more than $q+1$ points of $U$. Thus $\pi_\mathcal{B}$ meets $\Sigma_\infty$ in a line that does not belong to the spread $\mathcal{S}$, and so corresponds to an affine Baer subplane $\mathcal{B}$ of $PG(2,q^2)$.

Note that $\mathcal{B}$ contains at least one tangent line of $U$, namely, $\ell_2 = \ell_\infty$. Suppose that $\mathcal{B}$ contains another line $m'$ that when extended to $PG(2,q^2)$ is a tangent of $U$, in which case the point of tangency is $X = m' \cap m$ as $m \subset U$. Thus in $PG(4,q)$, $m'$ is a line that meets the set $\mathcal{U}$ (corresponding to the unital $U$) in the point $X$, and $X \neq V$. Let $m' \cap \Sigma_\infty = W$, let $m_W$ be the line of $\mathcal{S}$ through $W$, and let the corresponding point of $\ell_\infty$ be $Y = m' \cap \ell_\infty$. As $m_W \neq m_{P_2}$, necessarily $Y$ and $P_2$ are distinct points of $\mathcal{B} \cap \ell_\infty$. As $Y \in \ell_2$, the feet of $Y$ are collinear by our hypothesis, and thus lie on $P_2 X$. Hence the $q+1$ points of the Baer subline $m$ are the feet of $Y$. In $PG(4,q)$, let $M_1 = WP_1 \cap m$, and consider the plane $\pi = \langle m_W, WP_1 \rangle$. Then $\pi$ corresponds to a line of $PG(2,q^2)$ that contains the points $Y \in \ell_\infty$ and $M_1 \in m$, and hence this line is a tangent of $U$ as the points on $m$ are the feet of $Y$. But this line is not a tangent of $U$ since it also meets $U$ in $P_1$, a contradiction.

Therefore $\mathcal{B}$ contains exactly one line which when extended to $PG(2,q^2)$ is a tangent line of $U$. Hence by Theorem 6.1, $\mathcal{B}$ meets $U$ in precisely $2q+1$ points. We now show that the $q-1$ points of $\mathcal{B} \cap U$, other than $P_1$ and the points on $m$, all lie on the line $P_2 P_1$ of $\mathcal{B}$. Assume this is not the case, so that there is at least one point $P$ among these $q-1$ points which is not on the line $P_2 P_1$. Then in $PG(4,q)$, the plane $\pi_\mathcal{B}$ corresponding to $\mathcal{B}$ would contain the line $P_1 P$, which meets $m$ in a point $M_2$. The line $P_1 P$ of $\pi_{\mathcal{B}}$ corresponds to a Baer subline of $\mathcal{B}$ that meets $\ell_\infty$ and also contains the points $P_1$, $P$, and $M_2$ of $U$. As $P \notin P_1 P_2$, this line is the uniquely determined Baer subline that is the intersection of $P_1 P$ with $U$. However, $P_1 P \cap U$ does not contain a point of $\ell_\infty$ as $\ell_\infty$ is a tangent to $U$ at $P_2$. Hence we have a contradiction. Thus the secant line $P_1 P_2$ also meets $U$ in a Baer subline.

Therefore all the secants through $P_1$ and $P_2$ are Baer sublines. Hence by Theorem 7.6, $U$ is an ovoidal-Buekenhout-Metz unital with respect to $P_1$ and $P_2$, and then by Corollary 7.8 we know that $U$ is the classical unital.    $\square$

An interesting open question is whether the feet of all exterior points on *one* tangent line being collinear is enough to characterize the unital as ovoidal-Buekenhout-Metz.

### 7.3.3 Characterizations Using O'Nan Configurations

In [174] O'Nan shows that the classical unital does not contain four distinct lines meeting in six distinct points. This configuration, as previously discussed, is called an O'Nan configuration and is illustrated once again in Figure 7.10. (This configuration is also called the Pasch configuration in Design Theory literature.)

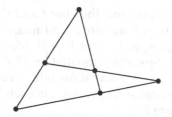

**Fig. 7.10.** An O'Nan configuration

A longstanding conjecture is that this property characterizes the classical unital; that is, a unital with no O'Nan configurations must be classical. Wilbrink [223] investigated this conjecture and proved a weaker characterization. He showed that if a unital of order $q$, $q$ even, has no O'Nan configurations and satisfies condition II (defined below), then $U$ is classical. He also introduced a third condition to characterize the classical unital of odd order. We briefly discuss this approach, and refer the reader to [223] for details.

We consider the unital as a design, not necessarily embedded in a projective plane. Thus $U$ is a 2-$(q^3 + 1, q + 1, 1)$ design. However, we nonetheless call the blocks of $U$ "lines". Wilbrink's first two geometric conditions on $U$ are the following:

I. $U$ contains no O'Nan configurations.

II. Suppose $U$ contains the following configuration: a line $\ell$, a point $P$ not on $\ell$, a line $m$ through $P$ meeting $\ell$, and a point $Q \neq P$ on $m$ that is distinct from $m \cap \ell$. Then necessarily there exists a line $\ell' \neq m$ through $Q$ which intersects all lines through $P$ that meet $\ell$ (see Figure 7.11).

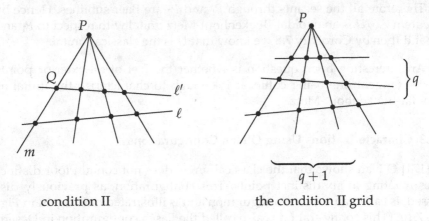

condition II                    the condition II grid

**Fig. 7.11.** Condition II

Note that condition I implies that the lines $\ell$ and $\ell'$ of condition II have no common points. In fact, these conditions lead to an equivalence relation on the lines of $U$ not through a given point $P$ of $U$. Namely, two lines $\ell, \ell'$ of $U$ not through $P$ are called **P-parallel**, denoted by $\ell \|_p \ell'$, if they meet the same lines through $P$. Hence, if $U$ is a unital for which conditions I and II hold, then $\|_p$ is an equivalence relation with $q$ lines in each $\|_p$-equivalence class, as shown in the grid of Figure 7.11.

Working in the Bruck-Bose representation, it is not difficult to show that the classical unital satisfies conditions I and II.

**Lemma 7.42.** *Let $U$ be a classical unital in $\mathrm{PG}(2, q^2)$. Then $U$ satisfies conditions I and II.*

*Proof.* Suppose $U$ contains an O'Nan configuration containing a point $P$. Let $\ell$ be the tangent to $U$ through $P$, and work in the Bruck-Bose representation with $\ell_\infty = \ell$. Thus $U$ corresponds to an elliptic cone $\mathcal{U}$ in $\mathrm{PG}(4, q)$ that meets $\Sigma_\infty$ in the spread line $p$ and has vertex $V$ on $p$. The given O'Nan configuration through $P$ corresponds to two lines $m_1, m_2$ of $\mathcal{U}$ through $V$ and two planes $\pi_1, \pi_2$ that meet $\Sigma_\infty$ in different lines of the spread, such that $\pi_1$ and $\pi_2$ both meet $m_1$ and $m_2$, and $\pi_1 \cap \pi_2 = X \in \mathcal{U}$. Then the point $X$ lies in $\langle \pi_1, m_1 \rangle \cap \langle \pi_2, m_2 \rangle = \langle m_1, m_2 \rangle$. Hence the plane $\langle m_1, m_2 \rangle$ contains more than $2q + 2$ points of $\mathcal{U}$, a contradiction as $\mathcal{U}$ is an elliptic cone. Therefore

$U$ satisfies condition I. Note that this proof shows that if $U$ is an ovoidal-Buekenhout-Metz unital with respect to the point $P_\infty = U \cap \ell_\infty$, then $U$ contains no O'Nan configurations with $P_\infty$ as a vertex.

Now, let $\ell$ be a line of $U$ that meets $q + 1$ lines of $U$ through $P$, and let $m$ be a line of $U$ through $P$ that meets $\ell$. Let $Q$ be a point of $U$ on $m$ that is distinct from $P$ and distinct from $m \cap \ell$. Then in $PG(4, q)$, $\ell$ corresponds to a plane $\pi$ that meets $q + 1$ lines of $\mathcal{U}$ through $V$. Thus the 3-space $\Sigma = \langle \pi, V \rangle$ meets the elliptic cone $\mathcal{U}$ in a quadratic cone. The plane $\pi$ contains a unique line $t$ of the spread, and $t$ is the only line of the spread in $\Sigma$. The plane $\langle t, Q \rangle$ corresponds to the unique line of $PG(2, q^2)$ that contains $Q$ and also meets the same $q + 1$ lines through $P$ as does $\ell$. That is, condition II holds for any classical unital $U$.                                                                $\square$

Wilbrink shows that if an arbitrary unital satisfies conditions I and II, then the incidence structure constructed in the same way as in Theorem 6.16 is the point residual of an inversive plane. He then uses this inversive plane structure to show that a *generalized quadrangle* (see [177] for an introduction to generalized quadrangles) can be constructed from any unital that satisfies conditions I and II. By working with this generalized quadrangle, Wilbrink is able to show for even $q$ that the converse of Lemma 7.42 is true. We state this highly nontrivial result here without proof.

**Theorem 7.43.** *Let $U$ be a unital of order $q$ satisfying conditions I and II. If $q$ is even, then $U$ is classical. In particular, $U$ is embedded in the Desarguesian plane $PG(2, q^2)$.*

In order to give a characterization involving O'Nan configurations for classical unitals of odd order, Wilbrink needed a third geometric condition:

III. Let $P$ be a point of $U$, and let $m_1, m_2, m_3$ be three distinct lines through $P$. Let $Q_i, R_i$ be two distinct points of $m_i \setminus P$, for $i = 1, 2, 3$. If $Q_1 Q_2 \|_p R_1 R_2$ and $Q_1 Q_3 \|_p R_1 R_3$, then $Q_2 Q_3 \|_p R_2 R_3$.

As before, working in the Bruck-Bose representation, one may show that the classical unital satisfies condition III. Using a group theoretic approach (basically, showing that the automorphism group of $U$ contains $PSU(3, q^2)$), Wilbrink was able to prove the following result.

**Theorem 7.44.** *If $U$ is a unital satisfying conditions I, II, and III, then $U$ is classical. In particular, $U$ is embedded in the Desarguesian plane $PG(2, q^2)$.*

Whether or not condition I by itself implies that $U$ is classical remains a longstanding open problem.

## 7.4 Characterizations Using the Quadratic Extension PG$(4, q^2)$

We now look at another characterization of the classical unital. In the Bruck-Bose representation of PG$(2, q^2)$ in PG$(4, q)$, let $S$ be the regular spread of $\Sigma_\infty$. Consider the quadratic extension PG$(4, q^2)$ of PG$(4, q)$. By Theorem 3.10, $S$ has two transversal lines $g$ and $g^\sigma$ in the quadratic extension PG$(3, q^2)$ of $\Sigma_\infty$. If $U$ is an orthogonal-Buekenhout-Metz unital in PG$(2, q^2)$, then $U$ corresponds to an elliptic cone $\mathcal{U}$ in PG$(4, q)$. This elliptic cone is given by a quadratic form $f(x) = 0$. In the quadratic extension PG$(4, q^2)$, the points satisfying the equation $f(x) = 0$ form a hyperbolic cone $\overline{\mathcal{U}}$ (that is, a cone whose vertex is a point and whose base is a three-dimensional hyperbolic quadric). Metsch [171] showed that $U$ is classical if and only if $\overline{\mathcal{U}}$ contains the transversal lines $g$ and $g^\sigma$ of $S$. Metsch used a coordinate argument to prove this result. Here we present the geometric proof given by Casse and Quinn in [89].

**Theorem 7.45.** *Let $U$ be an orthogonal-Buekenhout-Metz unital in* PG$(2, q^2)$, *so that $U$ corresponds to an elliptic cone $\mathcal{U}$ in* PG$(4, q)$ *given by the quadratic form $f(x) = 0$. Let $g$ be a transversal line of the regular spread $S$ of $\Sigma_\infty$. Then $U$ is classical if and only if in the quadratic extension* PG$(4, q^2)$, *$g$ lies on the hyperbolic cone $\overline{\mathcal{U}}$ corresponding to $f$.*

*Proof.* If $U$ is the classical unital, then every secant line of $U$ meets $U$ in a Baer subline. Hence in PG$(4, q)$, the elliptic cone $\mathcal{U}$ contains many Baer conics. Thus by Corollary 3.24, in PG$(4, q^2)$ the hyperbolic cone $\overline{\mathcal{U}}$ contains more than two distinct points of the transversal line $g$. Hence $g$ (and similarly $g^\sigma$) is a line of $\overline{\mathcal{U}}$.

Conversely, suppose that the hyperbolic cone $\overline{\mathcal{U}}$ in PG$(4, q^2)$ contains the transversal line $g$ of the regular spread $S$. As $\overline{\mathcal{U}}$ is the quadratic extension of the elliptic cone $\mathcal{U}$ of PG$(4, q)$, it follows that $\overline{\mathcal{U}}$ also contains the conjugate line $g^\sigma$ of $g$. Let $\pi$ be a plane of PG$(4, q)$ containing a spread element $m$ distinct from $p_\infty$. Then $\pi$ meets $\mathcal{U}$ in an irreducible conic $\mathcal{C}$, implying that the points $\overline{m} \cap g$ and $\overline{m} \cap g^\sigma$ are in the quadratic extension $\overline{\mathcal{C}}$ of $\mathcal{C}$. Hence by Corollary 3.24, $\mathcal{C}$ is a Baer conic. Therefore $U$ is an orthogonal-Buekenhout-Metz unital with respect to $P_\infty$ with a secant line not through $P_\infty$ meeting $U$ in a Baer subline. Thus by Theorem 7.2, $U$ is classical. $\square$

It is possible that working in PG$(4, q^2)$ may lead to new characterizations of classical and/or ovoidal-Buekenhout-Metz unitals.

## 7.5 The Bose Representation of PG$(2, q^2)$ in PG$(5, q)$

This section contains a short discussion of the Bose representation of the Desarguesian plane PG$(2, q^2)$ in PG$(5, q)$. We show that this setting leads to a characterization of the classical unital using the quadratic extension $PG(5, q^2)$. We provide relatively few details, referring the reader to [61] for a complete description of the Bose representation.

In the Bose representation, the points of $PG(2, q^2)$ correspond to lines of $PG(5, q)$ that form a line spread $\mathcal{S}$. In this setting, Thas [207] showed that the classical unital corresponds to a nonsingular elliptic quadric. The line spread $\mathcal{S}$ has two conjugate **transversal planes** in $PG(5, q^2)$. In his Ph.D. thesis, Maddock [167] studied this representation extensively. In particular, he obtained the following characterization of the classical unital by using the quadratic extension $PG(5, q^2)$.

**Theorem 7.46.** *An elliptic quadric $\mathcal{Q}$ of $PG(5, q)$ corresponds to a classical unital in $PG(5, q^2)$ via the Bose representation if and only if in the quadratic extension $PG(5, q^2)$, $\overline{\mathcal{Q}}$ contains the transversal planes of the spread $\mathcal{S}$.*

As the Bruck-Bose representation is obtained by intersecting the Bose representation with a hyperplane, we obtain a number of corollaries in the Bruck-Bose setting. One corollary of the above result is Theorem 7.45 (hence Maddock gives a geometric proof of Theorem 7.45 without using the characterization of Theorem 7.2). Another corollary is that a unital of $PG(2, q^2)$ corresponding to a nonsingular quadric is classical if and only if the quadric contains the transversal lines $g$ and $g^\sigma$ of the regular spread. This second corollary leads to a one-line proof of the characterization given in Theorem 7.5 that a nonsingular-Buekenhout unital is classical.

Maddock is able to use this setting to provide short proofs of several other known characterizations of the classical unital. As an example, he uses the quadratic extension to give the following very short proof of the characterization given by Lefèvre-Percsy in Theorem 7.2 (note that Maddock's proof of Theorem 7.45 does not use Lefèvre-Percsy's characterization).

**Theorem 7.2** *Let $U$ be an orthogonal-Buekenhout-Metz unital with respect to $P_\infty$. If $U$ has a secant line $\ell$ not containing $P_\infty$ which meets $U$ in a Baer subline, then $U$ is a classical unital.*

*Proof.* We use the notation of Section 7.4. The line $\ell$ meeting $U$ in a Baer subline corresponds in $PG(4, q)$ to a plane meeting $\mathcal{U}$ in a Baer conic $\mathcal{C}$. By Corollary 3.24, the quadratic extension $\overline{\mathcal{C}}$ of $\mathcal{C}$ meets the transversals $g$ and $g^\sigma$ of the regular spread $\mathcal{S}$ in points $P$ and $P^\sigma$, respectively. The hyperplane $\Sigma_\infty$ meets $\mathcal{U}$ in the line $p_\infty$ of $\mathcal{S}$, so that $\Sigma_\infty$ is a tangent hyperplane of the elliptic cone $\mathcal{U}$. Hence in the quadratic extension, the line $g$ meets $\overline{\mathcal{U}}$ twice at $g \cap \overline{p}_\infty$ and once at $P$. Hence, counted according to multiplicities, $g$ has three points on $\overline{\mathcal{U}}$ and thus $g$ is contained in $\overline{\mathcal{U}}$. Similarly, $g^\sigma$ is contained in $\overline{\mathcal{U}}$. Thus by Theorem 7.45, $U$ is classical. $\qquad\square$

Polverino [181] uses coordinates to show that an orthogonal-Buekenhout-Metz unital $U$ is represented in $PG(5, q)$ by an algebraic hypersurface of degree four minus the complement of a line in a three-dimensional subspace.

Furthermore, such a hypersurface is reducible over some extension of $GF(q)$ if and only if $U$ is classical.

Again, it seems possible to us that studying unitals in the Bose setting may lead to further characterizations of classical and/or ovoidal-Buekenhout-Metz unitals.

## 7.6 Group Theoretic Characterizations

There are many characterizations of unitals using group theory. We have decided not to discuss these in detail in this book, but have summarized most known results in Appendix B.

# 8

# Open Problems

In this short concluding chapter we present some open problems involving unitals. It is our hope that some of these problems will soon be solved, perhaps by someone reading this monograph. We list the problems in no particular order. Some problems appear to be much harder than others, and some have been mentioned previously in the text.

1. Find some unital embedded in the Desarguesian plane $PG(2, q^2)$ which is not a Buekenhout unital, or prove that none such exist.
2. Find a unital which can be embedded in both the Desarguesian plane and some non-Desarguesian plane, or show that none such exist.
3. Is there a coding theoretic characterization for orthogonal-Buekenhout-Metz unitals embedded in $PG(2, q^2)$? What about Buekenhout-Tits unitals embedded in $PG(2, q^2)$?
4. Determine the full automorphism group of the orthogonal-Buekenhout-Metz unitals embedded in $PG(2, q^2)$, treated as designs. Repeat for the Buekenhout-Tits unitals, treated as designs, which are embedded in Desarguesian planes.
5. Solve the isomorphism problem for orthogonal-Buekenhout-Metz unitals embedded in $PG(2, q^2)$, treated as designs.
6. Must any two Buekenhout-Tits unitals embedded in some Desarguesian plane be projectively equivalent? Must they be isomorphic as designs?
7. Find a projective plane of order $n^2$ which does not contain a unital of order $n$, if possible.
8. Determine the possible geometric configurations for the feet of an ovoidal-Buekenhout-Metz unital embedded in a Desarguesian plane.
9. Let $U$ be a unital embedded in the Desarguesian plane $PG(2, q^2)$. Suppose there is a tangent line $\ell$ to $U$ such that the feet of every point $Q \in \ell \setminus U$ are collinear. Must $U$ be an ovoidal-Buekenhout-Metz unital?
10. Does the absence of O'Nan configurations characterize the classical unital?

S. Barwick, G. Ebert, *Unitals in Projective Planes*,
DOI: 10.1007/978-0-387-76366-8_8, © Springer Science+Business Media, LLC 2008

11. It can be shown that no orthogonal-Buekenhout-Metz unital embedded in the Desarguesian plane $PG(2, q^2)$ contains an O'Nan configuration with $P_\infty$ as a vertex, although other O'Nan configurations are known to occur for small values of $q$. Is there a combinatorial characterization of orthogonal-Buekenhout-Metz unitals involving the absence of certain O'Nan configurations?

12. Let $U$ be a nonclassical unital in $PG(2, q^2)$ that is ovoidal-Buekenhout-Metz with respect to $P_\infty = U \cap \ell_\infty$. Let $\ell$ be a secant line through $P_\infty$, and let $D$ be a derivation set of $\ell$ with $D \neq U \cap \ell$. Determine whether or not deriving with respect to $D$ gives a unital in the Hall plane $Hall(q^2)$.

13. Find all (mutually inequivalent) packings of $\mathcal{H}(2, 25)$, the Hermitian curve embedded in $PG(2, 25)$.

14. Do nonsingular-Buekenhout unitals embedded in derivable translation planes admit packings?

15. Do all self-dual, two-dimensional semifield planes admit unitary polarities?

16. Characterize the ovoidal-Buekenhout-Metz unitals embedded in non-Desarguesian, self-dual translation planes which can be expressed as the absolute points of some unitary polarity.

17. Determine the dimensions over various prime fields of the linear codes generated by the characteristic vectors of the blocks of ovoidal-Buekenhout-Metz unitals embedded in the Desarguesian plane.

18. Show that the intersection of a classical unital and an ovoidal-Buekenhout-Metz unital embedded in $PG(2, q^2)$ has size congruent to 1 modulo $q$.

19. Determine the parameters of the codes arising from the ovoidal-Buekenhout-Metz unitals embedded in $PG(2, q^2)$, treated as curves, via the method of Goppa. Are these codes close to "maximum distance separable", as defined in coding theory?

20. A unital which admits a packing is often called *resolvable*. Show that a resolvable unital embedded in a two-dimensional translation plane is necessarily an ovoidal-Buekenhout-Metz unital.

21. Completely determine the stabilizers of the Ganley unitals embedded in the commutative Dickson semifield planes.

22. Determine if the unitals in the Albert twisted field planes, constructed by Abatangelo, Korchmáros, and Larato as a union of ovals, are actually orthogonal-Buekenhout-Metz unitals.

23. Let $P$ be a point on a unital $U$. Let $\mathcal{I}$ be the incidence structure whose *points* are the lines of $U$ through $P$. If $m$ is a line of $U$ not through $P$, then let $B_m$ be the set of lines through $P$ that meet $m$. Define the *blocks* of $\mathcal{I}$ to be the *distinct* sets $B_m$ as $m$ varies. Incidence in $\mathcal{I}$ is defined by containment in $U$. If $\mathcal{I}$ is the point residual of an inversive plane, is $U$ necessarily embedded in a projective plane? If $U$ is embedded in a translation plane of dimension at most two over its kernel and $\mathcal{I}$ is the point residual of an inversive plane, is $U$ necessarily an ovoidal-Buekenhout-Metz unital?

24. Use the quadratic extension $PG(4, q^2)$ of the Bruck-Bose representation to develop some new characterizations of the classical unital and/or ovoidal-Buekenhout-Metz unitals embedded in $PG(2, q^2)$.

25. Use the Bose representation in $PG(5, q)$ to develop some new characterizations of the classical unital and/or Buekenhout-Metz unitals embedded in $PG(2, q^2)$.

# A

# Nomenclature of Unitals

There has been an assortment of conflicting names for the unitals constructed by Buekenhout [81]. In this book we have tried to develop a standard naming system for the different classes of unitals. We have included this appendix to emphasize the names we have used. Let $\mathcal{P}$ be a translation plane of order $q^2$ with kernel containing $GF(q)$, and consider the Bruck-Bose representation of $\mathcal{P}$ in $PG(4, q)$.

Throughout this book we use the names below.

- The **classical** or **Hermitian** unital corresponds to a nondegenerate Hermitian curve in $PG(2, q^2)$.

- An **orthogonal-Buekenhout-Metz** unital in $\mathcal{P}$ corresponds to an orthogonal cone (elliptic cone) in $PG(4, q)$; that is, an ovoidal cone with an elliptic quadric as base.

- A **Buekenhout-Tits** unital in $\mathcal{P}$ corresponds to an ovoidal cone in $PG(4, q)$ with base a Tits ovoid, where $q \geq 8$ is an odd power of 2.

- An **ovoidal-Buekenhout-Metz** unital in $\mathcal{P}$ corresponds to an ovoidal cone in $PG(4, q)$ with *any* ovoid as base. So the class of ovoidal-Buekenhout-Metz unitals includes the orthogonal-Buekenhout-Metz unitals and the Buekenhout-Tits unitals.

- A **nonsingular-Buekenhout** unital in $\mathcal{P}$ corresponds to a nonsingular parabolic quadric in $PG(4, q)$.

- A **Buekenhout** unital in $\mathcal{P}$ corresponds in $PG(4, q)$ to either a nonsingular parabolic quadric or to an ovoidal cone with *any* ovoid as base. That is, it is a unital arising from either one of Buekenhout's two constructions.

In the literature, several different names have been used. The term *Hermitian arc* refers to a unital embedded in a projective plane (for example, see [192]). Buekenhout used the names *parabolic unital* and *hyperbolic unital* to refer to unitals that were, respectively, tangent or secant to the line at infinity. These names have been used throughout the literature, and various authors use *parabolic Buekenhout unital* and *hyperbolic Buekenhout unital* to refer to an ovoidal-Buekenhout-Metz and a nonsingular-Buekenhout unital, respectively. Lefèvre-Percsy [163] first used the name *Buekenhout-Metz unital* to refer to a unital that corresponds to an ovoidal cone. Many subsequent researchers also used this term. Other authors (for example, see [108]) have used *Buekenhout-Metz* to refer to unitals which have an elliptic quadric as base. However, Metz's proof of the existence of nonclassical unitals is not specific to the case where the cone is an elliptic cone. The name *Buekenhout-Tits* was first used in [109] to specify those unitals whose base is a Tits ovoid. The term *orthogonal Buekenhout-Metz* has been used by authors who want a name for the class of unitals corresponding to orthogonal cones (for example, see [39], [89]). Finally, in [33] the unitals corresponding to a nonsingular quadric are called *Buekenhout unitals*.

# B

## Group Theoretic Characterizations of Unitals

In this appendix we give a brief summary of the main theorems in articles that give group theoretic characterizations of unitals, and also summarize related group theoretic results.

**1971 Kantor [150].** Let $\mathcal{P}$ be a projective plane of order $q^2$, and let $U$ be a unital obtained as the absolute points of some unitary polarity of $\mathcal{P}$. Let $G(U)$ be the group of collineations stabilizing $U$. If $G$ is a subgroup of $G(U)$ that acts transitively on the flags $(X, \ell)$, where $X$ is a point of $U$ and $\ell$ is a secant line of $U$ through $X$, then $\mathcal{P}$ is Desarguesian and $\mathrm{PSU}(3, q) \leq G$. Moreover, if there are at least three noncollinear points $X \in U$ such that $G(U)$ contains $q$ $(X, \ell)$-elations (where $\ell$ is the pole of $X$ with respect to the polarity of $U$), then $\mathcal{P}$ is Desarguesian. In 1984 Camina and Gagen [83] noted that transitivity of $G$ on secants of $U$ implies flag-transitivity of $G$, thus improving the first result above.

**1972 Hoffer [137].** If $\mathcal{P}$ is a projective plane of order $q^2$ and $G$ is a collineation group of $\mathcal{P}$ that is isomorphic to $\mathrm{PSU}(3, q)$, then $\mathcal{P}$ is Desarguesian and $G$ contains all elations of $\mathcal{P}$ that commute with a suitable unitary polarity.

**1982 Biscarini [55].** Let $G$ be a group of collineations acting on a unital $U$ embedded in $\mathrm{PG}(2, q^2)$. If $G$ is transitive on the secant lines of $U$ and $G$ is generated by involutions, then $U$ is classical.

**1984 Abatangelo [1].** A Buekenhout-Metz unital $U$ of $\mathrm{PG}(2, q^2)$, $q$ odd, is classical if and only if there is a cyclic linear collineation group of order $q^2 - 1$ that stabilizes $U$ and fixes two distinct points of $U$. Applying this characterization, a new proof of Metz's result [172] is given.

**1985 Kantor [151].** The classification of all finite simple groups is used to characterize designs with 2-transitive automorphism groups. An application to unitals shows that the only unitals with an automorphism group that is 2-transitive on points are the classical unitals and the Ree unitals.

**1989 Biliotti and Korchmáros [51].** Let $G$ be a collineation group acting on a unital $U$ embedded in a projective plane $\mathcal{P}$ of order $q^2$, for some odd prime power $q \geq 5$. If $G$ is transitive on the points of $U$ and the socle of $G$ has even order, then $\mathcal{P}$ is Desarguesian, $U$ is classical, and $G \cong \mathrm{PSU}(3,q^2)$. A corollary of this result shows that if $G$ acts primitively on the points of $U$, then $\mathcal{P}$ is Desarguesian, $U$ is classical, and (for $q > 3$) $\mathrm{PSU}(3,q^2) \leq G \leq \mathrm{P\Gamma U}(3,q^2)$.

**1989 Biliotti and Korchmáros [52].** Let $G$ be a collineation group acting on a unital $U$ embedded in a projective plane $\mathcal{P}$ of even order $q^2$. If $G$ is transitive on the points of $U$ and the socle of $G$ has even order, then $G$ contains involutory elations, and hence by [150] $\mathcal{P}$ is Desarguesian, $U$ is classical, and $\mathrm{PSU}(3,q^2) \leq G \leq \mathrm{P\Gamma U}(3,q^2)$. As a consequence, if $G$ acts primitively on the points of $U$, then $\mathcal{P}$ is Desarguesian, $U$ is classical, and (for $q > 2$) $\mathrm{PSU}(3,q^2) \leq G \leq \mathrm{P\Gamma U}(3,q^2)$.

**1989 Doyen [107].** An investigation is conducted of 2-$(v,k,1)$ designs with an automorphism group that satisfies certain transitivity assumptions. An application to unitals shows that the only line-transitive unitals are the classical unitals and the Ree unitals.

**1991 Batten [40].** Let $S$ be a blocking set in a finite projective plane $\mathcal{P}$ of order $q^2$, and let $G$ be a collineation group of $\mathcal{P}$ that acts flag-transitively on $S$. Then $\mathcal{P} = \mathrm{PG}(2,q^2)$, $S$ is either a Baer subplane of $\mathcal{P}$ or a classical unital of $\mathcal{P}$, and $\mathrm{PSU}(3,q) \triangleleft G$.

**1991 Abatangelo and Larato [5].** Let $U$ be the nonclassical unital in $\mathrm{PG}(2,q^2)$, $q$ odd, constructed from a pencil of $q$ suitable conics (see [21] or [134]). Then the projective group $G$ stabilizing $U$ has order $2q^3(q-1)$ and is the semidirect product of a normal elementary abelian subgroup of order $q^3$ and a cyclic group of order $2(q-1)$. Conversely, a unital whose projective stabilizer has the above structure is necessarily such a pencil of conics. These results were obtained independently in [22].

**1992 Abatangelo [2].** Let $U$ be a nonclassical orthogonal-Buekenhout-Metz unital with respect to some point $P$ in $\mathrm{PG}(2,q^2)$, $q$ even. Then the collineation group $G$ stabilizing $U$ fixes $P$, has a normal subgroup acting sharply transitively on the points of $U \setminus P$, and has point stabilizers of order $q-1$ for any point of $U \setminus P$. This property characterizes the orthogonal-Buekenhout-Metz unitals among nonclassical unitals in $\mathrm{PG}(2,q^2)$, $q$ even. The first result was obtained independently in [108].

**1996 Ebert and Wantz [112].** A unital $U$ embedded in $\mathrm{PG}(2,q^2)$ is an orthogonal-Buekenhout-Metz unital if and only if $U$ admits a linear collineation group that is the semidirect product of a subgroup of order $q^3$ by a subgroup of order $q-1$. This is the full linear collineation group stabilizing $U$ except when $U$ is a classical unital or the union of a partial pencil of conics (as constructed in [21] or [134]). As a corollary, one obtains the result that a unital $U$ embedded in $\mathrm{PG}(2,q^2)$ is classical if and only if $U$ admits a linear collineation group that is the semidirect product of a subgroup of order $q^3$ by a subgroup of order $q^2 - 1$.

**1996 Abatangelo and Larato [6].** Let $U$ be an ovoidal-Buekenhout-Metz unital embedded in $PG(2, q^2)$, $q$ even, and let $G$ be the linear collineation group stabilizing $U$. Then $U$ is an orthogonal-Buekenhout-Metz unital if and only if there is a point $P$ of $U$ such that the stabilizer of $P$ in $G$ has a subgroup that acts sharply transitively on $U \setminus P$. Furthermore, if $U$ is a Buekenhout-Tits unital in $PG(2, 2^r)$, where $r \geq 3$ is an odd integer, then the full linear collineation group leaving $U$ invariant is an abelian group of order $q^2$ that fixes some point of $U$. The second result was independently obtained in [109].

**1999 Abatangelo, Enea, Korchmáros, and Larato [3].** The collineation stabilizer is completely determined for the unital obtained as the set of absolute points for the natural unitary polarity of a commutative twisted field plane of odd order $q^2$.

**1999 Biliotti [48].** A study is conducted of the structure of collineation groups which preserve a unital in a finite projective plane of order $q^2$, where $q \equiv 1 \pmod 4$. In particular, several results are obtained when the group is a 2-group.

**2000 Cossidente, Ebert, and Korchmáros [91].** A unital in $PG(2, q^2)$ is classical if and only if it is stabilized by a Singer subgroup of $PGL(3, q^2)$ of order $q^2 - q + 1$.

**2001 Cossidente, Ebert, and Korchmáros [92].** A unital in $PG(2, q^2)$ is classical if and only if it is stabilized by a linear collineation group of order $6(q + 1)^2$ which fixes neither a point nor a line in $PG(2, q^2)$.

**2001 Abatangelo, Korchmáros, and Larato [4].** Let $U$ be a unital embedded in a translation plane $\mathcal{P}$ of odd order so that there is a collineation group $G$ stabilizing $U$ which fixes the point $P = U \cap \ell_\infty$ and acts transitively on the points of $U \setminus P$. If $G$ contains an affine homology, then $\mathcal{P}$ is a semifield plane and $G$ has a normal subgroup $K$ that acts sharply transitively on $U \setminus P$. In addition, if $K$ is abelian, then $\mathcal{P}$ is a commutative semifield plane.

**2002 Giuzzi [123].** A new proof of the group theoretic characterization by Cossidente, Ebert, and Korchmáros [91] is given.

**2002 Abatangelo and Larato [7].** Let $\mathcal{P}$ be a commutative Dickson semifield plane of odd order $q^2$, and suppose that $U$ is a transitive parabolic unital (see [4]) obtained as the absolute points of a unitary polarity of $\mathcal{P}$. Then the sharply transitive normal subgroup $K$ of the collineation stabilizer of $U$ is non-abelian.

**2006 Jha and Johnson [145].** Let $U$ be a unital embedded in a translation plane $\mathcal{P}$ of order $q^2$ so that there is a collineation group $G$ stabilizing $U$ which fixes the point $P = U \cap \ell_\infty$ and acts transitively on the points of $U \setminus P$. Assume that $q$ is a power of the prime $p$ and $G$ has order $q^3 u$, where $u$ is a prime $p$-primitive divisor of $q^2 - 1$. Then $\mathcal{P}$ is Desarguesian. Other group theoretic results for such planes also are obtained.

**2006 Johnson [146].** Let $\mathcal{P}$ be a semifield plane of order $q^2$ that is two-dimensional over its kernel. Then $\mathcal{P}$ admits a transitive parabolic unital; that is, a unital $U$ with a collineation group $G$ that fixes the point $P = U \cap \ell_\infty$ and acts transitively on the points of $U \backslash P$).

# References

1. L.M. Abatangelo. Una caratterizzazione gruppale delle curve Hermitiane. *Le Matematiche (Catania)*, **39** (1984) 101–110.
2. V. Abatangelo. On Buekenhout-Metz unitals in PG$(2, q^2)$, $q$ even. *Arch. Math.*, **59** (1992) 197–203.
3. V. Abatangelo, M.R. Enea, G. Korchmáros, and B. Larato. Ovals and unitals in commutative twisted field planes. Combinatorics (Assisi, 1996), *Discrete Math.*, **208/209** (1999) 3–8.
4. V. Abatangelo, G. Korchmáros, and B. Larato. Transitive parabolic unitals in translation planes of odd order. *Discrete Math.*, **231** (2001) 3–10.
5. V. Abatangelo and B. Larato. A group theoretic characterization of parabolic Buekenhout-Metz unitals. *Boll. U.M.I.*, **5A** (7) (1991) 195–206.
6. V. Abatangelo and B. Larato. A characterization of Buekenhout-Metz unitals in PG$(2, q^2)$, $q$ even. *Geom. Dedicata*, **59** (1996) 137–145.
7. V. Abatangelo and B. Larato. Polarity and transitive parabolic unitals in translation planes of odd order. *J. Geometry*, **74** (2002) 1–6.
8. V. Abatangelo, B. Larato, and L.A. Rosati. Unitals in planes derived from Hughes planes. *J. Combin. Inf. Syst. Sci.*, **15** (1990) 151–155.
9. A. Aguglia. Designs arising from Buekenhout-Metz unitals. *J. Combin. Des.*, **11** (2003) 79–88.
10. A. Aguglia and G.L. Ebert. A combinatorial characterization of classical unitals. *Arch. Math.*, **78** (2002) 166–172.
11. A.A. Albert. On non-associative division algebras. *Trans. Amer. Math. Soc.*, **72** (1952) 296–309.
12. J. André. Über nicht-Desarguessche Ebenen mit transitiver Translationgruppe. *Math. Z.*, **60** (1954) 156–186.
13. B.R. Andriamanalimanana. *Ovals, Unitals and Codes*. Ph.D. thesis, Lehigh University, 1979.
14. E.F. Assmus Jr. and J.D. Key. Arcs and ovals in the Hermitian and Ree unitals. *Eur. J. Combin.*, **10** (1989) 297–308.
15. E.F. Assmus Jr. and J.D. Key. Baer subplanes, ovals and unitals. *IMA Math. Appl. Coding Theory and Design Theory, Part I: Coding Theory*, **20** (1990) 1–8.
16. E.F. Assmus Jr. and J.D. Key. *Designs and Their Codes*. Cambridge University Press, 1992.

17. R. Baer. Polarities in finite projective planes. *Bull. Amer. Math. Soc.*, **52** (1946) 77–93.

18. R. Baer. *Linear Algebra and Projective Geometry*. Dover, 2005 (Academic Press, 1952).

19. B. Bagchi and S. Bagchi. Designs from pairs of finite fields. I. A cyclic unital $U(6)$ and other regular Steiner 2-designs. *J. Combin. Theory Ser. A*, **52** (1989) 51–61.

20. R.D. Baker. Polarities of elliptic semiplanes. *Proc. Ninth Southeast. Conf. Combin. Graph Theory Comp.*, (1978) 85–96.

21. R.D. Baker and G.L. Ebert. Intersections of unitals in the Desarguesian plane. *Congr. Numer.*, **70** (1990) 87–94.

22. R.D. Baker and G.L. Ebert. On Buekenhout-Metz unitals of odd order. *J. Combin. Theory Ser. A*, **60** (1992) 67–84.

23. R.D. Baker, G.L. Ebert, G. Korchmáros, and T. Szőnyi. Orthogonally divergent spreads of Hermitian curves. Finite Geometry and Combinatorics (Deinze, 1992), *London Math. Soc. Lecture Note Ser.*, **191** (1993) 17–30.

24. R.D. Baker and K.L. Wantz. Unitals in the code of the Hughes plane. *J. Combin. Des.*, **12** (2004) 35–38.

25. S. Ball. Polynomials in finite geometries. Surveys in Combinatorics (Canterbury, 1999), *London Math. Soc. Lecture Note Ser.*, **267** (1999) 17–35.

26. S. Ball, A. Blokhuis, and C.M. O'Keefe. On unitals with many Baer sublines. *Des. Codes Cryptogr.*, **17** (1999) 237–252.

27. S. Ball and M.R. Brown. The six semifield planes associated with a semifield flock. *Adv. Math.*, **189** (2004) 68–87.

28. A. Barlotti. Un'estensione del teorema di Segre-Kustaanheimo. *Boll. U.M.I.*, **10** (1955) 498–506.

29. A. Barlotti. Some topics in finite geometric structures. *Institute of Statistics Mimeo Series No. 439*, University of North Carolina, Chapel Hill, 1965.

30. A. Barlotti. Finite geometries and designs. Surveys in Combinatorics, *London Math. Soc. Lecture Note Ser.*, **124** (1987) 1–11.

31. A. Barlotti and G. Lunardon. Una classe di unitals nei Δ-piani. *Riv. Mat. Univ. Parma*, (4) **5** (1979) 781–785.

32. M. Barnabei and F. Bonetti. Two examples of finite Bolyai-Lobachevsky planes. *Rend. Mat.*, **12** (1979) 291–296.

33. S.G. Barwick. A characterization of the classical unital. *Geom. Dedicata*, **52** (1994) 175–180.

34. S.G. Barwick. A class of Buekenhout unitals in the Hall plane. *Bull. Belg. Math. Soc. Simon Stevin*, **3** (1996) 113–124.

35. S.G. Barwick. Unitals in the Hall plane. *J. Geometry*, **58** (1997) 26–42.

36. S.G. Barwick, L.R.A. Casse, and C.T. Quinn. The André/Bruck and Bose representation in $PG(2h, q)$: unitals and Baer subplanes. *Bull. Belg. Math. Soc. Simon Stevin*, **7** (2000) 173–197.

37. S.G. Barwick and C.M. O'Keefe. Unitals and inversive planes. *J. Geometry*, **58** (1997) 43–52.

38. S.G. Barwick, C.M. O'Keefe, and L. Storme. Unitals which meet Baer subplanes in 1 modulo $q$ points. *J. Geometry*, **68** (2000) 16–22.

39. S.G. Barwick and C.T. Quinn. Generalising a characterisation of Hermitian curves. *J. Geometry*, **70** (2001) 1–7.

40. L.M. Batten. Blocking sets with flag transitive collineation groups. *Arch. Math.*, **56** (1991) 412–416.

41. C. Bernasconi and R. Vincenti. Spreads induced by varieties $\mathcal{V}_2^3$ of $PG(4, q)$ and Baer subplanes. *Boll. Un. Mat. Ital. B*, **18** (1981) 821–830.

42. E. Bertini. Geometria proiettiva degli iperspasi. E. Spoerri, ed., Pisa, 1907.

43. A. Betten, D. Betten, and V.D. Tonchev. Unitals and codes. *Discrete Math.*, **267** (2003) 23–33.

44. A. Beutelspacher. Embedding the complement of a Baer subplane or a unital in a finite projective plane. *Mitt. Math. Sem. Giessen*, **163** (1984) 189–202.

45. A. Beutelspacher. Embedding linear spaces with two line degrees in finite projective planes. *J. Geometry*, **26** (1986) 43–61.

46. A. Beutelspacher and U. Rosenbaum. *Projective Geometry: From Foundations to Applications*. Cambridge University Press, 1998.

47. J. Bierbrauer. *Introduction to Coding Theory*. Chapman & Hall/CRC, 2005.

48. M. Biliotti. Collineation groups preserving a unital in a projective plane of order $m^2$ with $m \equiv 1(4)$. *J. London Math. Soc.*, **60** (1999) 589–606.

49. M. Biliotti, V. Jha, and N.L. Johnson. *Foundations of Translation Planes*. Marcel Dekker, 2001.

50. M. Biliotti, V. Jha, N.L. Johnson, and A. Montinaro. Two-transitive groups on a hyperbolic unital. Preprint.

51. M. Biliotti and G. Korchmáros. Collineation groups preserving a unital of a projective plane of odd order. *J. Algebra*, **122** (1989) 130–149.

52. M. Biliotti and G. Korchmáros. Collineation groups preserving a unital of a projective plane of even order. *Geom. Dedicata*, **31** (1989) 333–344.

53. M. Biliotti and G. Lunardon. Insiemi di derivazione e sottopiani di Baer. *Atti Accad. Lincei, Rend. Sci. Fis. Mat. Nat.*, Serie VIIIa, **LXIX** (1980) 135–141.

54. G. Birkhoff and J. von Neumann. The logic of quantum mechanics. *Ann. Math.*, **37** (1936) 823–843.

55. P. Biscarini. Hermitian arcs of $PG(2, q^2)$ with a transitive collineation group on the set of $(q+1)$–secants. *Rend. Sem. Mat. Brescia*, **7** (1982) 111–124.

56. A. Blokhuis, A. Brouwer, and H. Wilbrink. Hermitian unitals are codewords. *Discrete Math.*, **97** (1991) 63–68.

57. A. Blokhuis and C.M. O'Keefe. On deriving unitals. 16th British Combinatorial Conference (London, 1997), *Discrete Math.*, **197/198** (1999) 137–141.

58. E. Boros and T. Szőnyi. On the sharpness of a theorem of B. Segre. *Combinatorica*, **6** (1986) 261–268.

59. R.C. Bose. On the construction of balanced incomplete block designs. *Ann. Eugenics*, **9** (1939) 353–399.

60. R.C. Bose. Mathematical theory of symmetrical factorial design. *Sankhyā*, **8** (1947) 107–166.

61. R.C. Bose. On a representation of the Baer subplanes of the Desarguesian plane $PG(2, q^2)$ in a projective five dimensional space $PG(5, q)$. *Proceedings of the International Colloquium on Combinatorial Theory*, Rome (1973) 381–391.

62. R.C. Bose, R.C. Freeman, and D.G. Glynn. On the intersection of two Baer subplanes in a finite projective plane. *Utilitas Math.*, **17** (1980) 65–77.

63. P.L.H. Brooke. On the Steiner system $S(2, 4, 28)$ and codes associated with the simple group of order 6048. *J. Algebra*, **97** (1985) 376–406.

64. A.E. Brouwer. Some unitals on 28 points and their embeddings in projective planes of order 9. Geometries and Groups, *Springer Lecture Notes*, **893** (1981) 183–188.

65. A.E. Brouwer. Some unitals on 28 points and their embeddings in projective planes of order 9. *Math. Centre Report ZW155181*, Amsterdam, 1981.

66. A.E. Brouwer. A unital in the Hughes plane of order nine. *Discrete Math.*, **77** (1989) 55–56.

67. A.E. Brouwer, W.H. Haemers, and H.A. Wilbrink. Some 2-ranks. *Discrete Math.*, **106/107** (1992) 83–92.

68. M.R. Brown. Ovoids of $PG(3, q)$, $q$ even, with a conic section. *J. London Math. Soc.*, **62** (2000) 569–582.

69. R.H. Bruck. Difference sets in a finite group. *Trans. Amer. Math. Soc.*, **78** (1955) 464–481.

70. R.H. Bruck. Quadratic extensions of cyclic planes. *Proc. Symp. Appl. Math.*, **10** (1960) 15–44.

71. R.H. Bruck. Construction problems of finite projective planes. *Conference on Combinatorial Mathematics and its Applications*, University of North Carolina Press, (1969) 426–514.

72. R.H. Bruck. Circle geometry in higher dimensions. *A Survey of Combinatorial Theory*. North Holland, (1973) 69–77.

73. R.H. Bruck and R.C. Bose. The construction of translation planes from projective spaces. *J. Algebra*, **1** (1964) 85–102.

74. R.H. Bruck and R.C. Bose. Linear representations of projective planes in projective spaces. *J. Algebra*, **4** (1966) 117–172.

75. A. Bruen. Baer subplanes and blocking sets. *Bull. Amer. Math. Soc.*, **76** (1970) 342–344.

76. A. Bruen. Blocking sets in finite projective planes. *SIAM J. Appl. Math.*, **21** (1971) 380–392.

77. A.A. Bruen and J.C. Fisher. An observation on certain point-line configurations in classical planes. *Discrete Math.*, **106/107** (1992) 93–96.

78. A.A. Bruen and J.W.P. Hirschfeld. Intersections in projective spaces I: Combinatorics. *Math. Z.*, **193** (1986) 215–225.

79. A.A. Bruen and R. Silverman. Arcs and blocking sets II. *Eur. J. Combin.*, **8** (1987) 351–356.

80. A.A. Bruen and J.A. Thas. Blocking sets. *Geom. Dedicata*, **6** (1977) 193–203.

81. F. Buekenhout. Existence of unitals in finite translation planes of order $q^2$ with a kernel of order $q$. *Geom. Dedicata*, **5** (1976) 189–194.

82. F. Buekenhout. Characterizations of semi quadrics. *Atti Conv. Lincei*, **17** (1976) 393–421.

83. A.R. Camina and T.M. Gagen. Block transitive automorphism groups and designs. *J. Algebra*, **86** (1984) 549–554.

84. J. Cannon and C. Playoust. *An Introduction to MAGMA*. University of Sydney Press, 1993.

85. L.R.A. Casse. *A Guide to Projective Geometry*. Oxford University Press, 2006.

86. L.R.A. Casse and G. Lunardon. On Buekenhout-Metz unitals. Finite Geometry and Combinatorics (Deinze, 1997), *Bull. Belg. Math. Soc. Simon Stevin*, **5** (1998) 237–240.

87. L.R.A. Casse, C.M. O'Keefe, and T. Penttila. Characterisations of Buekenhout-Metz unitals. *Geom. Dedicata*, **59** (1996) 29–42.

88. L.R.A. Casse and C.T. Quinn. The Bruck-Bose map and Buekenhout-Metz unitals. R. C. Bose Memorial Conference (Fort Collins, CO, 1995), *J. Stat. Plann. Inference*, **72** (1998) 89–98.

89. L.R.A. Casse and C.T. Quinn. Ruled cubic surfaces in $PG(4, q)$, Baer subplanes of $PG(2, q^2)$ and Hermitian curves. *Discrete Math.*, **248** (2002) 17–25.

90. H.S.M. Coxeter. *Projective Geometry, Second Edition.* Springer-Verlag, 1987.

91. A. Cossidente, G.L. Ebert, and G. Korchmáros. A group-theoretic characterization of classical unitals. *Arch. Math.*, **74** (2000) 1–5.

92. A. Cossidente, G.L. Ebert, and G. Korchmáros. Unitals in finite Desarguesian planes. *J. Algebraic Combin.*, **14** (2001) 119–125.

93. M. deFinis. On *k*-sets of type $(m, n)$ in projective planes of square order. Finite Geometries and Designs, *London Math. Soc. Lecture Note Ser.*, **49** (1981) 98–103.

94. M.J. de Resmini. There exists at least three non isomorphic $S(2, 4, 28)$'s. *J. Geometry*, **16** (1981) 148–151.

95. M.J. de Resmini and N. Hamilton. Hyperovals and unitals in Figueroa planes. *Eur. J. Combin.*, **19** (1998) 215–220.

96. P. Dembowski. *Finite Geometries.* Springer-Verlag, 1968 (reprinted 1997).

97. L.J. Dickey. Embedding the complement of a unital in a projective plane. *Atti Conv. Geom. Appl.*, Perugia, (1971) 199–203.

98. L.E. Dickson. *Linear Groups with an Exposition of the Galois Field Theory.* Teubner, 1901 (Dover, 1958).

99. L.E. Dickson. On finite algebras. *Nachr. kgl. Ges. Wiss. Göttingen*, (1905) 358–393.

100. G. Donati and N. Durante. Some subsets of the Hermitian curve. *Eur. J. Combin.*, **24** (2003) 211–218.

101. G. Donati and N. Durante. On the intersection of Hermitian curves and of Hermitian surfaces. *Discrete Math.*, in press.

102. J. Dover. Some design-theoretic properties of Buekenhout unitals. *J. Combin. Des.*, **4** (1996) 449–456.

103. J.M. Dover. A family of non-Buekenhout unitals in the Hall planes. Mostly Finite Geometries (Iowa City, IA, 1996), *Lecture Notes Pure Appl. Math.*, **190** (1997) 197–205.

104. J.M. Dover. Spreads and resolutions of Ree unitals. *Ars Combin.*, **54** (2000) 301–309.

105. J.M. Dover. Subregular spreads of Hermitian unitals. *Des. Codes Cryptogr.*, **39** (2006) 5–15.

106. J.M. Dover. A search for spreads of Hermitian unitals. Private communication.

107. J. Doyen. Designs and automorphism groups. Surveys in Combinatorics, *London Math. Soc. Lecture Note Ser.*, **141** (1989) 74–83.

108. G.L. Ebert. On Buekenhout-Metz Unitals of even order. *Eur. J. Combin.*, **13** (1992) 109–117.

109. G.L. Ebert. Buekenhout-Tits unitals. *J. Algebraic Combin.*, **6** (1997) 133–140.

110. G.L. Ebert. Hermitian arcs. *Rend. Circ. Mat. Palermo*, **51** (1998) 87–105.

111. G.L. Ebert. Buekenhout unitals. Combinatorics (Assisi, 1996), *Discrete Math.*, **208/209** (1999) 247–260.

112. G.L. Ebert and K. Wantz. A group-theoretic characterization of Buekenhout-Metz unitals. *J. Combin. Des.*, **4** (1996) 143–152.

113. G. Faina. Sylow *p*-subgroups of $PSU(3, q^2)$. *Atti Sem. Mat. Fis. Univ. Modena*, **32** (1983) 255–259.

114. G. Faina. A characterization of the tangent lines to a Hermitian curve. *Rend. Mat.*, **3** (1983) 553–557.

115. G. Faina. Intrinsic characterizations of Hermitian curves. *Simon Stevin*, **65** (1991) 165–178.

116. G. Faina and G. Korchmáros. A graphic characterization of Hermitian curves. *Ann. Discrete Math.*, **18** (1983) 335–342.

117. R. Figueroa. A family of not $(V, \ell)$-transitive projective planes of order $q^3$, $q \not\equiv 1$ (mod 3) and $q > 2$. *Math. Z.*, **181** (1982) 471–479.

118. J.C. Fisher, J.W.P. Hirschfeld, and J.A. Thas. Complete arcs in planes of square order. *Ann. Discrete Math.*, **30** (1986) 243–250.

119. J.W. Freeman. Reguli and pseudo–reguli in $PG(3, s^2)$. *Geom. Dedicata*, **9** (1980) 267–280.

120. M.J. Ganley. Polarities in translation planes. *Geom. Dedicata*, **1** (1972) 28–40.

121. M.J. Ganley. A class of unitary block designs. *Math. Z.*, **128** (1972) 34–42.

122. L. Giuzzi. Collineation groups of the intersection of two classical unitals. *J. Combin. Des.*, **9** (2001) 445–459.

123. L. Giuzzi. A characterisation of classical unitals. *J. Geometry*, **74** (2002) 86–89.

124. V.D. Goppa. Algebraic-geometric codes. *Math. USSR-Izv.*, **21** (1983) 75–91.

125. T. Grundhöfer. A synthetic construction of Figueroa planes. *J. Geometry*, **26** (1986) 191–201.

126. K. Grüning. Das Kleinste Ree-Unital. *Arch. Math.*, **46** (1986) 473–480.

127. K. Grüning. A class of unitals of order $q$ which can be embedded in two different translation planes of order $q^2$. *J. Geometry*, **29** (1987) 61–77.

128. M. Hall. *Theory of Groups, Second Edition*. Chelsea Publishing Co., 1976.

129. R. Hill. *A First Course in Coding Theory*. Oxford University Press, 1986.

130. J.W.P. Hirschfeld. *Finite Projective Spaces of Three Dimensions*. Oxford University Press, 1985.

131. J.W.P. Hirschfeld. *Projective Geometry over Finite Fields, Second Edition*. Oxford University Press, 1998.

132. J.W.P. Hirschfeld, L. Storme, J.A. Thas, and J.F. Voloch. A characterization of Hermitian curves. *J. Geometry*, **41** (1991) 72–78.

133. J.W.P. Hirschfeld and T. Szőnyi. Sets in a finite plane with few intersection numbers and a distinguished point. *Discrete Math.*, **97** (1991) 229–242.

134. J.W.P. Hirschfeld and J.A. Thas. *General Galois Geometries*. Oxford University Press, 1991.

135. G. Hiss. On the incidence matrix of the Ree unital. *Des. Codes Cryptogr.*, **10** (1997) 57–62.

136. G. Hiss. Hermitian function fields, classical unitals, and representations of 3-dimensional unitary groups. *Indag. Math. (N.S.)*, **15** (2004) 223–243.

137. A.R. Hoffer. On unitary collineation groups. *J. Algebra*, **22** (1972) 211–218.

138. G. Hölz. Constructions of designs which contain a unital. *Arch. Math.*, **37** (1981) 179–183.

139. D.R. Hughes. A class of non-Desarguesian projective planes. *Canad. J. Math.*, **9** (1957) 378–388.

140. D.R. Hughes and F.C. Piper. *Projective Planes*. Springer-Verlag, 1973.

141. D.R. Hughes and F.C. Piper. *Design Theory*. Cambridge University Press, 1985.

142. W.A. Jackson, C.M. O'Keefe, and T. Penttila. On the unitals of Mathon, Bagchi and Bagchi. *Univ. West. Aust. Research Report 46*, University of Western Australia, 1990.

143. W.A. Jackson, F.C. Piper, and P.R. Wild. Non-embedding of non prime power unitals with point–regular groups. *Discrete Math.*, **98** (1991) 23–28.

144. D.B. Jaffe and V.D. Tonchev. Computing linear codes and unitals. *Des. Codes Cryptogr.*, **14** (1998) 39–52.

145. V. Jha and N.L. Johnson. Translation planes of order $q^2$ admitting collineation groups of order $q^3 u$ preserving a parabolic unital. *Note di Mat.*, **26** (2006) 119–137.

146. N.L. Johnson. Transitive parabolic unitals in semifield planes. *J. Geometry*, **85** (2006) 61–71.

147. N.L. Johnson, V. Jha, and M. Biliotti. *Handbook of Finite Translation Planes*. Chapman and Hall, 2007.

148. N.L. Johnson and G. Lunardon. On the Bose-Barlotti-planes. *Geom. Dedicata*, **49** (1994) 173–182.

149. N.L. Johnson and R. Pomareda. Collineation groups of translation planes admitting hyperbolic Buekenhout or parabolic Buekenhout-Metz unitals. *J. Combin. Theory Ser. A*, **114** (2007) 658–680.

150. W.M. Kantor. On unitary polarities of finite projective planes. *Canad. J. Math.*, **23** (1971) 1060–1077.

151. W.M. Kantor. Homogeneous designs and geometric lattices. *J. Combin. Theory Ser. A*, **38** (1985) 66–74.

152. B.C. Kestenband. Unital intersections in finite projective planes. *Geom. Dedicata*, **11** (1981) 107–117.

153. B.C. Kestenband. Degenerate unital intersections in finite projective planes. *Geom. Dedicata*, **13** (1982) 101–106.

154. B.C. Kestenband. A family of unitals in the Hughes plane. *Canad. J. Math.*, **42** (1990) 1067–1083.

155. B.C. Kestenband. New families of balanced incomplete block designs with block size $q + 1$. *J. Stat. Plann. Inference*, **30** (1992) 83–91.

156. J.D. Key. Some applications of Magma in designs and codes: oval designs, Hermitian unitals and generalized Reed-Muller codes. *J. Symbolic Comput.*, **31** (2001) 37–53.

157. J.D. Key and N.K.A. Rostom. Unitary designs with regular sets of points. *Discrete Math.*, **69** (1988) 235–239.

158. P.B. Kleidman. The finite flag transitive linear spaces with an exceptional automorphism group. Finite Geometries and Combinatorial Designs, *Contemp. Math.*, **111** (1987) 117–136.

159. G. Korchmáros. Inherited arcs in finite affine planes. *J. Combin. Theory Ser. A*, **42** (1986) 140–143.

160. B. Larato. Una caraterizzazione degli unitals parabolici di Buekenhout-Metz. *Le Matematiche (Catania)*, **38** (1983) 95–98.

161. B. Larato. Strutture d'incidenza associate a curve hermitiane. *Boll. U.M.I.*, **3A** (1984) 87–95.

162. C. Lefèvre. Tallini sets in projective spaces. *Atti Accad. Naz. Lincei Rend.*, **59** (1975) 392–400.

163. C. Lefèvre-Percsy. Characterization of Buekenhout-Metz unitals. *Arch. Math.*, **36** (1981) 565–568.

164. C. Lefèvre-Percsy. Characterization of Hermitian curves. *Arch. Math.*, **39** (1982) 476–480.

165. K.H. Leung and Q. Xiang. On the dimensions of the binary codes of a class of unitals. *Discrete Math.*, to appear.

166. H. Lüneburg. Some remarks concerning the Ree groups of type $(G_2)$. *J. Algebra*, **3** (1966) 256–259.

167. R. Maddock. *The Bose Representation of* $PG(2, q^2)$ *and Some Associated Varieties*. Ph.D. thesis, University of Adelaide, 2000.

168. G. McGuire, V.D. Tonchev, and H.N. Ward. Characterizing the Hermitian and Ree unitals on 28 points. *Des. Codes Cryptogr.*, **13** (1998) 57–61.

184     References

169. R. Mathon. Constructions for cyclic Steiner 2-designs. *Ann. Discrete Math.*, **34** (1987) 353–362.
170. R. Mathon and T. van Trung. Unitals and unitary polarities in symmetric designs. *Des. Codes Cryptogr.*, **10** (1997) 237–250.
171. K. Metsch. A note on Buekenhout-Metz unitals. Geometry, Combinatorial Designs and Related Structures (Spetses, 1996), *London Math. Soc. Lecture Note Ser.*, **245** (1997) 177–180.
172. R. Metz. On a class of unitals. *Geom. Dedicata*, **8** (1979) 125–126.
173. B. Mortimer. The modular permutation representations of the known doubly transitive groups. *Proc. London Math. Soc.*, **41** (1980) 1–20.
174. M.E. O'Nan. Automorphisms of unitary block designs. *J. Algebra*, **20** (1972) 495–511.
175. W.F. Orr. *The Miquelian Inversive Plane IP(q) and the Associated Projective Planes.* Ph.D. thesis, University of Wisconsin - Madison, 1973.
176. G. Panella. Caratterizzazione delle quadriche di uno spazio (tri-dimensionale) lineare sopra un corpo finito. *Boll. Un. Mat. Ital.*, **10** (1955) 507–513.
177. S.E. Payne and J.A. Thas. *Finite Generalized Quadrangles.* Pitman, 1984.
178. T. Penttila and G.F. Royle. Sets of type $(m, n)$ in the affine and projective planes of order nine. *Des. Codes Cryptogr.*, **6** (1995) 229–245.
179. F.C. Piper. Polarities in the Hughes plane. *Bull. London Math. Soc.*, **2** (1970) 209–213.
180. F.C. Piper. Unitary block designs. Graph Theory and Combinatorics, *Res. Notes Math.*, **34** (1979) 98–105.
181. O. Polverino. Linear representation of Buekenhout-Metz unitals. *Discrete Math.*, **267** (2003) 247–252.
182. C.T. Quinn. *Baer Structures, Unitals and Associated Finite Geometries.* Ph.D. thesis, University of Adelaide, 1996.
183. C.T. Quinn and L.R.A. Casse. Concerning a characterisation of Buekenhout-Metz unitals. *J. Geometry*, **52** (1995) 159–167.
184. B. Qvist. Some remarks concerning curves of the second degree in a finite plane. *Ann. Acad. Sci. Fenn. Ser. A*, **134** (1952).
185. A. Rahilly. Embeddings of generalized unitals and semi-regular polarities. *Ars Combin.*, **23** (1987) 273–302.
186. M. Richardson. On finite projective games. *Proc. Amer. Math. Soc.*, **7** (1956) 453–465.
187. G. Rinaldi. Hyperbolic unitals in the Hall planes. *J. Geometry*, **54** (1995) 148–154.
188. G. Rinaldi. Constructions of unitals in the Hall planes. *Geom. Dedicata*, **56** (1995) 249–255.
189. B. Rokowska. A new construction of the block systems $B(4, 1, 25)$ and $B(4, 1, 28)$. *Coll. Math.*, **38** (1977) 165–167.
190. L.A. Rosati. I gruppi di collineazioni dei piani di Hughes. *Boll. U.M.I.*, **13** (1958) 505–513.
191. L.A. Rosati. Designi unitari nei piani di Hughes. *Geom. Dedicata*, **27** (1988) 295–299.
192. M. Tallini Scafati. $\{k; n\}$-archi di un piano grafico finito, con particolare riguardo a quelli con due caratteri. *Atti Acad. Naz. Lincei Rend.*, **40** (1966) Nota I, 812–818, Nota II, 1020–1025.
193. M. Tallini Scafati. Caratterizzazione grafica delle forme hermitiane di un $S_{r,q}$. *Rend. Mat. Appl.*, **26** (1967) 273–303.

194. B. Segre. Curve razionali normali e $k$-archi negli spazi finiti. *Ann. Mat. Pura Appl.*, **39** (1955) 357–379.

195. B. Segre. Ovals in a finite projective plane. *Canad. J. Math.*, **7** (1955) 414–416.

196. B. Segre. On complete caps and ovaloids in three-dimensional Galois spaces of characteristic two. *Acta Arith.*, **5** (1959) 315–322.

197. B. Segre. Teoria de Galois, fibrazioni proiettive e geometrie non-desarguesiane. *Ann. Mat. Pura Appl.*, **64** (1964) 1–76.

198. B. Segre. Forme e geometrie hermitiane con particolare riguardo al caso finito. *Ann. Mat.*, **70** (1965) 1–201.

199. B. Segre. Introduction to Galois geometries. *Atti Accad. Naz. Lincei Mem.*, **8** (1967) 133–236. (edited by J.W.P. Hirschfeld.)

200. M. Seib. Unitäre Polaritäten endlicher projectiver Ebenen. *Arch. Math.*, **21** (1970) 103–112.

201. J.G. Semple and L. Roth. *Introduction to Algebraic Geometry.* Oxford University Press, 1949.

202. M.A. Shokrollahi. Codes on Hermitian curves. *Lecture Notes Comput. Sci.*, **307** (1988) 168–176.

203. S.D. Stoichev and V. Tonchev. Unital designs in planes of order 16. Coding, Cryptography and Computer Security (Lethbridge, 1998), *Discrete Appl. Math.*, **102** (2000) 151–158.

204. M. Sved. On configurations of Baer subplanes of the projective plane over a finite field of square order. *Lecture Notes Math.*, **952** (1982) 423–443.

205. G. Tallini. Sulle $k$-calotte uno spazio lineare finito. *Ann. Mat. Pura Appl.*, **42** (1956) 119–164.

206. D.E. Taylor. Unitary block designs. *J. Combin. Theory Ser. A*, **16** (1974) 51–56.

207. J.A. Thas. Semipartial geometries and spreads of classical polar spaces. *J. Combin. Theory Ser. A*, **35** (1983) 58–66.

208. J.A. Thas. Solution of a classical problem on finite inversive planes. *Finite Geometries, Buildings and Related Topics* (Pingree Park, CO, 1988), 145–159.

209. J.A. Thas. A combinatorial characterization of Hermitian curves. *J. Algebraic Combin.*, **1** (1992) 97–102.

210. J.A. Thas, P.J. Cameron, and A. Blokhuis. On a generalization of a theorem of B. Segre. *Geom. Dedicata*, **43** (1992) 299–305.

211. H.J. Tiersma. Remarks on codes from Hermitian curves. *IEEE Trans. Inf. Theory*, **33** (1987) 605–609.

212. J. Tits. Ovoides et groupes du Suzuki. *Arch. Math.*, **13** (1962) 187–198.

213. V.D. Tonchev. Unitals in the Hölz design on 28 points. *Geom. Dedicata*, **38** (1991) 357–363.

214. J.H. Van Lint and T.A. Springer. Generalized Reed-Solomon codes from algebraic geometry. *IEEE Trans. Inf. Theory*, **33** (1987) 305–309.

215. O. Veblen and J.H.M. Wedderburn. Non Desarguesian and non Pascalian geometries. *Trans. Amer. Math. Soc.*, **8** (1907) 379–380.

216. R. Vincenti. Fibrazioni di un $S_{3,q}$ indotte da fibrazioni di un $S_{3,q^2}$ e rappresentazione di sottopiani di Baer di un piano proiettivo. *Atti Mem. Accad. Sci. Lett. Arti Modena*, Serie VI, **XIX** (1977) 1–8.

217. R. Vincenti. Alcuni tipi di varieta $V_2^3$ di $S_{4,q}$ e sottopiani di Baer. *Suppl. BUMI*, **2** (1980) 31–44.

218. R. Vincenti. Varieties representing Baer subplanes of a translation plane of the class V. *Atti Sem. Mat. Fis. Univ. Modena*, **XXIV** (1980) 48–59.

219. R. Vincenti. A survey on varieties of $PG(4, q)$ and Baer subplanes of translation planes. *Ann. Discrete Math.*, **18** (1983) 775–780.

220. J. von Neumann and O. Morgenstern. *Theory of Games and Economic Behavior, Second Edition*. Princeton University Press, 1947.

221. K.L. Wantz. A new class of unitals in the Hughes plane. *Geom. Dedicata*, **70** (1998) 125–138.

222. K.L. Wantz. Unitals in the regular nearfield planes. *J. Geometry*, to appear.

223. H. Wilbrink. A characterization of the classical unitals. Finite Geometries, *Lecture Notes Pure Appl. Math.* (Marcel Dekker), **82** (1983) 445-454.

# Notation Index

| | |
|---|---|
| $a, b, c, \ldots$ | elements in $\mathrm{GF}(q)$ |
| $\alpha, \beta, \gamma, \ldots$ | elements in $\mathrm{GF}(q^2)$ |
| $\sigma$ | the conjugate map $x \to x^q$ for $x \in \mathrm{GF}(q^2)$ |
| $\zeta$ | a primitive element in $\mathrm{GF}(q^2)$ with primitive polynomial $x^2 - t_1 x - t_0$ |
| $x'$ | transpose of $x$ |
| $X^*$ | dual space of $X$ |
| $\mathrm{PG}(2, q^2)$ | classical/field/Desarguesian plane of order $q^2$ |
| $\mathrm{Hall}(q^2)$ | Hall plane of order $q^2$ |
| $\mathrm{Hgh}(q^2)$ | Hughes plane of order $q^2$ |
| $\mathrm{Fig}(q^6)$ | Figueroa plane of order $q^6$ |
| $\mathcal{H} = \mathcal{H}(2, q^2)$ | classical unital in $\mathrm{PG}(2, q^2)$ |
| $\mathcal{A}$ | an affine plane |
| $\mathcal{D}(\mathcal{A})$ | the derived affine plane |
| $\mathcal{P}$ | a projective plane |
| $\mathcal{D}(\mathcal{P})$ | the derived projective plane |
| $\mathcal{S}$ | a spread in $\mathrm{PG}(3, q)$ |
| $\mathcal{P}(\mathcal{S})$ | the projective plane constructed from the spread $\mathcal{S}$ via the Bruck-Bose representation |
| $U$ | a unital in $\mathrm{PG}(2, q^2)$ |
| $\mathcal{U}$ | the corresponding point set in $\mathrm{PG}(4, q)$ |
| $\ell, m, \ldots$ | lines in a projective space |
| $\pi, \ldots$ | planes in a projective space |
| $\Sigma, \Pi, \ldots$ | 3-spaces in a projective space |

# Index

absolute, 17
  non, 17
affine
  geometry, 5
  plane, 1
Albert twisted field, 100
alternating form, 16
annihilator, 14
arc, 123
  complete, 123
automorphism, 2
axis, 33

Baer conic, 46
Baer involution, 27
Baer ruled cubic, 49
Baer subline
  classical, 25
  projective plane, 26
Baer subplane
  affine, 35
  classical, 26
  projective plane, 26
base
  quadric cone, 11
block, 28
blocking set, 30
  minimal, 30
  reduced, 30
Bose representation, 165
Bruck-Bose map, 54
Bruck-Bose representation, 42
  notation, 42

Buekenhout unital, 64, 171
Buekenhout-Tits unital, 62, 171

center (collineation), 33
central collineation, 33
  axis, 33
  center, 33
characteristic, 8
characteristic vector, 115
chord, 22
circle
  inversive plane, 121
classical
  affine geometry, 5
  affine plane, 4
  projective geometry, 5
  projective plane, 4
  unital, 29, 171
code (linear), 128
collinear, 1
collineation, 2, 33
  automorphic, 7
  central, 33
  nonlinear, 7
companion automorphism
  semilinear transformation, 7
  sesquilinear form, 13
concurrent, 1
cone
  conic, 10
  elliptic, 11
  orthogonal, 11
  ovoidal, 12

point, 11
  quadratic, 10
  quadric, 10
conic, 9
  exterior point, 9
  interior point, 9
  irreducible, 9
  nondegenerate, 9
  nucleus, 9
conic cone, 10
conjugation, 8
coordinates
  dual, 6
  homogeneous, 6
correlation, 15

degenerate
  Hermitian curve, 24
  quadric, 9
derivable plane, 35
derivation, 36
derivation set, 35
derived plane
  affine, 35
  projective, 36
Desarguesian, 5
design, 28
  block, 28
  point, 28
Dickson semifield, 98
dimension, 5
  projective, 3, 5
Dimension Theorem, 3
discriminant, 8
dual
  plane, 2
  space, 14
  unital, 28

elation, 33
elliptic cone, 11
elliptic quadric, 10
equivalent
  projectively, 19
  unital, 29
exceptional nearfield plane, 103

feet, 23, 152, 153
fixed linewise, 33

fixed pointwise, 33
form
  alternating, 16
  Hermitian, 16
  irreducible binary quadratic, 10
  sesquilinear, 13
  skew-symmetric bilinear, 16
  symmetric bilinear, 16
Frobenius automorphism, 8
Fundamental Theorem, 7

generator
  cone, 11
  elliptic cone, 11
  ovoidal cone, 13
  ruled cubic surface, 48
Gram matrix, 17

Hall
  plane, 34
  quasifield, 34
Hermitian curve
  degenerate, 24
  nondegenerate, 18
Hermitian form, 16
Hermitian unital, 171
Hermitian variety
  nondegenerate, 18
homogeneous coordinates, 6
homography, 7
homology, 33
hyperbolic line, 22
hyperbolic quadric, 10
hyperovals, 123
hyperplane, 5
  polar, 17
  tangent, 11

infinity
  line, 2
  point, 2
internal structure, 121
intersection of
  two unitals, 110, 113
  unital and Baer subplane, 109–112
inversive plane
  egg-like, 121
  Miquelian, 121
involutory, 8

irreducible
    binary quadratic form, 10
    conic, 9
    quadratic equation, 8
isomorphic
    design, 28
    plane, 2
    unital, 29
isomorphism, 2
isotropic, 17
    non, 17
    totally, 17

kernel, 34

line at infinity, 2
linear code, 128
    binary, 128
    dimension, 128
    length, 128
    minimum weight, 128

nearfield, 101
nondegenerate
    conic, 9
    Hermitian curve, 18
    Hermitian variety, 18
    quadric, 9
    sesquilinear form, 13
nonsingular
    quadric, 9
    transformation, 7
nonsingular-Buekenhout unital, 64, 171
norm, 8
nucleus, 9
null polarity, 16

O'Nan configuration, 87
order
    affine plane, 4
    projective plane, 4
    unital, 28
ordinary polarity, 16
orthogonal complement, 14
orthogonal cone, 11
orthogonal polarity, 16
orthogonal-Buekenhout-Metz unital, 64,
    171
oval, 12

ovoid, 12
    Suzuki-Tits, 12
ovoidal cone, 12
ovoidal-Buekenhout-Metz unital, 64, 171

packing, 117
parabolic quadric, 11
parallel, 1
    axiom, 1
    classes, 1
perspectivity, 33
plane
    André, 36
    classical, 4
    Desarguesian, 5
    field, 4
    Figueroa, 103
    Hall, 34, 36, 89
    Hughes, 105
    Moulton, 3
    nearfield, 101
    real projective, 3
    regular nearfield, 101
    semifield, 97
    translation, 34
point at infinity, 2
point cone, 11
point-residual, 122
polar hyperplane, 17
polarity, 16
    null, 16
    ordinary, 16
    orthogonal, 16
    pseudo, 16
    symplectic, 16
    unitary, 16
pole, 17
primitive element, 8
Principle of Duality, 2
projective
    completion, 3
    dimension, 3, 5
    equivalence, 19
    geometry, 4
    plane, 2
projectivity, 7
pseudopolarity, 16

quadratic cone, 10

quadric, 9
  degenerate, 9
  elliptic, 10
  hyperbolic, 10
  nondegenerate, 9
  nonsingular, 9
  parabolic, 11
  singular, 9
quadric cone, 10
  base, 11
  vertex, 11
quasifield, 34
  Hall, 34
quotient space, 63

reciprocity property, 152
reflexive, 15
regular spread, 37
  transversal, 40
regulus, 36
  opposite, 37
  reversal, 53
  reverse, 37
ruled cubic surface, 48
  Baer ruled cubic, 49
  conic directrix, 48
  generators, 48
  line directrix, 48

secant line
  Baer subplane, 109
  Baer subspace, 38
  Hermitian curve, 22
  Hermitian variety, 22
  unital, 109
secant plane
  elliptic quadric, 10
  ovoid, 121
self-dual, 29
  projective plane, 20
  unital, 29
semifield, 97
  Albert twisted field, 100
  Dickson, 98
sesquilinear form, 13
set of type $(0, 1, 2, q + 1)$, 136
set of type $(m, n)$, 148
Singer
  cycle, 124

group, 124
singular quadric, 9
skew-symmetric bilinear form, 16
slope point, 2
solid, 5
spread, 37
  design, 117
  orthogonally divergent, 119
  regular, 37
standard position, 25
Suzuki-Tits ovoid, 12
symmetric bilinear form, 16
symplectic polarity, 16

tangent hyperplane
  elliptic quadric, 11
  ovoidal cone, 13
tangent line
  Baer subplane, 109
  Hermitian variety, 22
  unital, 109
tangent plane
  elliptic cone, 11
  elliptic quadric, 10
  hyperbolic quadric, 10
  ovoid, 121
  ovoidal cone, 13
trace, 9
  absolute, 8
transformation
  linear, 7
  semilinear, 7
transitive, 33
translation, 33
translation line, 34
translation plane, 34
  kernel, 34
transversal (regular spread), 40
transversal planes, 165
twisted ladder, 48

unital, 28
  Buekenhout, 64, 171
  Buekenhout-Tits, 62, 171
  chord, 22
  classical, 29, 171
  dual, 28
  embedded, 28
  Hermitian, 171

nonsingular-Buekenhout, 64, 171
orthogonal-Buekenhout-Metz, 64, 171
ovoidal-Buekenhout-Metz, 64, 171
Ree, 29
self-dual, 29
unitary polarity

classical projective geometry, 16
projective plane, 20

vertex
quadric cone, 11

weight, 128